Steven H. Lonsdale
Creatures of Speech
Lion, Herding, and Hunting Similes in the Iliad

# Beiträge zur Altertumskunde

Herausgegeben von
Ernst Heitsch, Ludwig Koenen,
Reinhold Merkelbach, Clemens Zintzen

Band 5

Springer Fachmedien Wiesbaden GmbH

# Creatures of Speech
# Lion, Herding, and Hunting Similes in the Iliad

Steven H. Lonsdale

Springer Fachmedien Wiesbaden GmbH 1990

ISBN 978-3-663-12002-5        ISBN 978-3-663-12001-8 (eBook)
DOI 10.1007/978-3-663-12001-8


CIP-Titelaufnahme der Deutschen Bibliothek

**Lonsdale, Steven H.:**
Creatures of speech, lion, herding, and hunting similes in the
Iliad / Steven H. Lonsdale. – Stuttgart: Teubner, 1990
(Beiträge zur Altertumskunde; 5)
ISBN 978-3-663-12002-5
NE: GT

# MGN

## ΟΜΟΦΡΟΣΥΝΗΙ

# Contents


Works Cited by Author ............................................................ IX

Preface ............................................................................. XI

Introduction ........................................................................ 1

I  Animals in the Similes ....................................................... 9
    The Predominance and Functions of Animal Similes................ 9
    The Oral Nature, Date, and Distribution of Similes................. 13

II  Animals in the Poems........................................................ 19
    Traditional Themes and Attitudes Towards Animals................. 19
    Animal Species: Principles of Selection............................... 25
    Names, Gender, and Epithets for Animals........................... 28
    Interaction of Language in Describing Motion and Emotion....... 33

III  The Beast of War: Lion Similes........................................... 39
    Typology............................................................................ 39
    Anatomy............................................................................ 42
    Action in Lion Similes......................................................... 46

IV  The Lion as Marauder of the Herds and as Hunter ...................... 49
    The *Aristeia* of Diomedes: Book 5....................................... 50
    The *Aristeia* of Agamemnon: Book 11 ................................. 56
    Reversal: Hektor Storms the Wall: Book 12........................... 60
    Hektor Storms the Ships: Book 15 ....................................... 65
    Conclusion ....................................................................... 69

V  Pursuit in Battle: Hunting Similes ........................................ 71
    Characteristics and Affinities with Lion Similes..................... 71
    The Roles of the Hound and Boar ....................................... 74
    Hunter and Hounds: Hektor and the Trojans: Books 11 and 17 .. 76

VI  Pursuit and Attack: Reversals in Hunting Imagery ...................... 85
    Imagery of Pursuit: Books 18–21......................................... 86
    Imagery of Attack: Book 22 ............................................... 90

VII Conclusions: Animal Imagery in the Homeric Narrative ............. 103
    Interaction of Animal Simile and Other Contexts .................. 110
    Simile   and   Omen .............................................. 112
    Simile and Epiphany ............................................. 115
    Simile and Ecphrasis ............................................ 117
    Lion Similes and the Theme of the Cattle Raid .................... 118
    Simile and Paradigm ............................................. 122
    Conclusion ...................................................... 125
Appendixes
    A.  A Register of Identifiable Species in the Poems .............. 129
    B.  A List of Emotive States Attributed to Animals .............. 133
        Expressions and Words for Emotive States and Seats of
        Emotion  Applied  to  Animals  .............................. 133
    C. Formulae in Lion Similes ..................................... 137
        Tables Showing Phraseology from Lion Similes ............ 139
        Table 1: Corresponding Phrases Restricted to Lion
            Similes ................................................. 139
        Table 2: Non-Restricted Repeated Lion Vocabulary ... 140
        Table 3: Lion Vocabulary Shared with Other Animals . 141
    D. A List of Line References to Lion Similes in Homer .......... 143
Bibliography ....................................................... 145
Index Locorum ..................................................... 159

# Works Cited by Author


Bowra, *Tradition*=C. M. Bowra, *Tradition and Design in the Iliad* (Oxford 1930; reprint, Westport, Conn. 1977).

Fenik, Typical Battle Scenes=B. Fenik, *Typical Battle Scenes in the Iliad*, Hermes Einzelschrift 21 (Wiesbaden 1968).

Fränkel, *Gleichnisse*=H. Fränkel, *Die homerischen Gleichnisse². Mit einem Nachwort und einem Literaturverzeichnis*, ed. E. Heitsch (Göttingen 1977).

Hainsworth, *Flexibility*=J. B. Hainsworth, *The Flexibility of the Homeric Formula* (Oxford 1968).

Kirk, *Commentary*=G. S. Kirk, *The Iliad: A Commentary I: Books 1–4* (Cambridge 1985).

Kirk, *Songs*=G. S. Kirk, *The Songs of Homer* (Cambridge 1962).

*KlPauly=Der Kleine Pauly* (Stuttgart 1964–75).

Körner, *Homerische Tierwelt*=O. Körner, *Die homerische Tierwelt²* (Munich 1930).

Krischer, *Formale Konventionen*=T. Krischer, *Formale Konventionen der homerischen Epik*, Zetemata 56 (Munich 1971).

Leaf, *Commentary*=W. Leaf, *The Iliad*, 2 vols. (1900–02).

Lee, *Similes Compared*=D. Lee, *The Similes of the Iliad and Odyssey Compared* (Melbourne 1966).

LSJ=H. Liddell, R. Scott and H. S. Jones, *A Greek–English Lexicon⁹* (Oxford 1940).

Moulton, *Similes*=C. Moulton, *Similes in the Homeric Poems*, Hypomnemata 49 (Göttingen 1977).

Parry, *MH*=A. Parry, ed., *The Making of Homeric Verse. The Collected Papers of Milman Parry* (Oxford 1971).

*RE*=W. von Pauly, G. Wissowa, *Real–Encyclopädie der klassischen Altertumswissenschaft* (Stuttgart 1890–).

Redfield, Nature=J. M. Redfield, *Nature and Culture in the Iliad. The Tragedy of Hektor* (Chicago 1975).

Richter, *Landwirtschaft*=W. Richter, *Die Landwirtschaft im homerischen Zeitalter, Archaeologia Homerica* H (Göttingen 1968).

Schnapp–Gourbeillon, Lions=A. Schnapp–Gourbeillon, *Lions, héros, masques. Les représentations de l' animal chez Homère* (Paris 1981).

Scott, *Homeric Simile*=W. C. Scott, *The Oral Nature of the Homeric Simile, Mnemosyne* Supp. 28 (Leiden 1974).

Segal, *Theme*=C. Segal, *The Theme of the Mutilation of the Corpse, Mnemosyne* Supp. 27 (Leiden 1973).

Shipp, *Studies*=G. P. Shipp, *Studies in the Language of Homer*² (Cambridge 1972).

Stanford, Commentary=W. B. Stanford, *The Odyssey of Homer*, I: Books I–XII (London 1947); II: Books XIII–XXIV (London 1948).

# Preface

This monograph has its origins in a Cambridge Ph. D. dissertation directed by Geoffrey S. Kirk and Patricia E. Easterling, and examined by Nicholas J. Richardson and Frank H. Stubbings. The body of the work has since been revised, and the Introduction and the first two and last chapters have been added. A grant from the Office of the Vice President for Academic Affairs at Davidson College enabled me to prepare a camera–ready copy with the secretarial assistance of Cheryl Branz and the expert advice of John Heil in the Philosophy Department. I want to extend appreciation to my colleagues in the Department of Classical Studies, Dirk French and Michael K. Toumazou, for reading portions of the manuscript.

It is a special pleasure to record my long–standing debt to Pat Easterling for maieutic gifts so freely bestowed then and now. To Richard Janko I must also express my deep gratitude for his continuing encouragement and for cheerful commentary on the manuscript both as a dissertation and as a monograph. Lowell Edmunds and S. Douglas Olson read and made insightful observations on the opening chapters. I am grateful to Hoyt Rogers, who, with attention to content and important detail, exercised acute editorial judgment, and in so doing helped smooth the prose. Ludwig Koenen's circumspect questions and criticisms have proven valuable in preparing the final version of this work. Any errors or infelicities that remain are mine alone.

In the notes titles of articles are given only if they have a special significance. Full titles are listed in the bibliography.

In the analysis of similes distinction is made only between (1) short and (2) developed similes.[1] (1) Short similes are usually preceded by an introductory word such as ὡς, followed by a noun which may be modified by an adjective or participle, although an adverbial or participial equivalent of ὡς may follow the noun, e.g. συὶ εἴκελος ἀλκήν. There are two types

---

[1] The division of similes into four types by T. B. L. Webster, *From Mycenae to Homer* (1958) 224f., has not been adopted.

of longer similes. The first type (2a) is introduced by ὡς and contains a finite verb.[2] The second type (2b) consists of a separate sentence preceded by ὡς ὅτε, etc., and may include one or more relative clauses.[3] In developed similes "introductory clause" refers to the clause commencing with ὡς ὅτε, etc., and leading into the similes (Fränkel's "*Wiesatz*"); "resumptive clause" refers to the clause commencing with ὥς (Fränkel's "*Sosatz*") and continuing until a full or half stop. Among animal similes in the strict sense of those describing the activities of lion, boar, and the like, are included pastoral and hunting similes, even though human actors may figure prominently in them.

For extended analyses of similes I have sometimes given in parentheses or in the notes terms used by Krischer, *Formale Konventionen* 24, for the stages of *aristeiai*: *Einzelkämpfe, Ansturm auf die Phalangen, Verfolgung des Heeres, Wiederherstellung, Monomachie*, and *Kampf um die Leiche*.[4]

Unless otherwise noted, the line numbers of textual citations in Greek refer to the 1931 Allen edition of the *Iliad* and the 1962 Von der Mühll edition of the *Odyssey*. Arabic numbers alone refer to Books of the *Iliad* (1.1); Arabic numbers preceded by the abbreviation for the *Odyssey* refer to Books of the *Odyssey* (*Od.* 1.1). In line numbers "f." is used for reference through the following verse, and "ff." for three or more verses; the latter seems desirable since the exact point at which a simile concludes is sometimes ambiguous. Translations are from Richmond Lattimore, *The Iliad* (Chicago 1951), and *The Odyssey* (New York 1965).

---

[2] Called a "simile proper" by M. Coffey, "The Function of the Homeric Simile," *AJP* 78 (1957) 113f.

[3] Called a "long simile" by Coffey, *idem*.

[4] This omits the first two stages of Krischer's scheme, arming and the radiance of arms, which are not accompanied by animal similes; see below, p. 13, n. 18.

# Introduction

Anyone who reads the *Iliad* is immediately struck by the prevalence of lion imagery in the poem, especially in the developed similes describing the behavior of the hero. The lion, essentially identical with the war hero, is the animal simile *par excellence*: the king of beasts and the king of similes. The lion is to the similes what the horse is to the narrative: a foil to the warrior but also an actor in his own right.[1] In the miniaturist world of the similes the lion, like the courageous warrior, is admired for his strength, agility, swift attack, persistence, and ingenuity. Like the aggressors encamped for nearly a decade outside Troy, the beast of the similes is an ever–present menace to the shepherd and his sheepflocks—a marauder who threatens to disrupt the tranquility of the entire community. But in the face of vigilant men and dogs, the lion can also find himself in a vulnerable position, turn cowardly, and flee. Several times the lion is the potential victim, and in a memorable simile from Book 18, a huntsman robs the lioness of her whelps. Lion similes are not all honorific. Apollo compares Achilles' excesses in Book 24 to a lion endowed with fierceness in his "manly heart" (ἀγήνορι θυμῷ, 24.42) and rushing against the sheepflocks to devour his meal (δαῖτα, 43), a word which has sinister implications in context.[2] The Homeric portrait of the lion therefore diverges from the stereotype of the kingly beast passed on from the Near East.[3] The Homeric oral tradition has refashioned the lion in the image of the hero and has thereby left an indelible imprint on the poem.[4]

The lion, however, is only one of several actors in the animal similes, which are populated by farmers, their dogs and livestock on the one hand,

---

[1] Delebecque, *Cheval* 75, observes that the horse is an omnipresent actor in the narrative.

[2] Cf. C. W. Macleod, *Homer. Iliad XXIV* (1982), *ad loc*.

[3] The question of lions in Asia Minor is discussed on pp. 103–106.

[4] J. Boardman, *Ancient Greek Art and Iconography*, ed. W. G. Moon (1983) 29, and others have argued that the characteristic Homeric lion simile may represent the signature of the poet.

and hunters, their hounds and quarry, on the other. An *agon* in the pastoral
and natural world distantly resounds with the martial conflict at Troy.

What is the relationship between the lion simile, the master simile, and
hunting and herding imagery? This study argues that the lion simile
orchestrates and integrates the hunting and herding imagery, as it gathers
up other types of metaphors in a subtle network that underlies the poem. In
order to demonstrate this assertion, this work focuses on the lion similes in
the war poem (with particular attention to the *aristeiai* of Diomedes,
Agamemnon, Hektor, and Achilles), their kinship with herding and hunting similes, and their crucial role in shaping the narrative. Textual analyses
concentrate on the principal type of lion simile—what may be called the
marauding lion simile—in which the feline harries the herdsman. Marauding lion similes account for over half of all developed lion similes, the most
frequent simile type in the poems.[5] But since the narrator's point of view
characteristically shifts from describing the psychology of the animal attacker to that of the human defenders, these similes can be regarded as either lion or herding similes. We feel drawn to identify with the victorious
beast but also experience a sympathy toward the shepherd and his flocks.
The fundamental relationship of marauding lion similes to the pastoral
background of the poem, and in particular to the theme of cattle–theft,
makes them cumulatively serve as a unified metaphor of the efforts of the
Achaians to storm the walls of the Trojan citadel and plunder the possessions within. The phraseology used to describe the conflict between man
and animal in the marauding lion simile serves in turn as the model for
sketching the clash between hunter and boar in the boar–hunting similes,
where the lion in fact is sometimes presented as an alternate subject. A typology for the marauding lion simile is established, using the simile at
20.164ff. as a model on which virtually all other lion, herding, and hunting
similes are based.

Analysis of the text of similes necessarily involves examining the
context. I am interested in assessing the interaction of simile and narrative
found, for example, in the vulture simile at 16.428–30:[6]

---

[5] About 55%, or 82 of the 146 verses comprising Homeric lion similes; see Ap–
pendix D for a list of line references to marauding and other lion similes.

[6] M. Baltes, *A&A* 29 (1983) 36-48, demonstrates how this simile, as well as the
simile of two contending lions (16.756ff.) and the famous lion-boar simile (16.823ff.),
not only interact significantly with their immediate context but progressively increase in
length and degree of overlap with the narrative to put into relief the central event in the
book, the death of Patroklos that prompts Achilles' return to the fighting.

οἳ δ᾽ ὥς τ᾽ αἰγυπιοὶ γαμψώνυχες ἀγκυλοχεῖλαι
πέτρηι ἐφ᾽ ὑψηλῆι μεγάλα κλάζοντε μάχωνται,
ὣς οἳ κεκλήγοντες ἐπ᾽ ἀλλήλοισιν ὄρουσαν.

The repetition of the verb κλάζω in the forms κλάζοντε and κεκλή-γοντες is the type of linguistic interaction between simile and narrative that characterizes the epic simile.[7] κλάζω is equally appropriate to the screaming of birds and to the shouting of warriors. In this example another case of interaction is to be found in μάχωνται (429). Although it actually describes the wrangling of vultures, μάχομαι is properly a military term, and its use here in the animal simile resonates significantly with the human objects of comparison in the narrative. In the bull simile occurring at Sarpedon's death (16.487ff.), βεβρυχώς for the hero's "roaring" in the narrative (486) is better suited to the bull than, say, στενάχων, which is actually used of the bull in the simile (489).[8] Linguistic interaction is symptomatic of the conceptual nature underlying the animal simile. In furnishing an analogue in the animal world, the poet creates a hybrid that reflects the bestial and human aspects of human nature.

The study is concerned, then, with such metaphorical uses of anthropomorphic vocabulary in the animal similes, as well as the opposite: words and phrases in the vicinity of an animal simile that are properly theriomorphic. The interaction between animal simile and narrative often involves the sharing of descriptive vocabulary, or the swapping of what is normally human for what is normally animal. This mixture of language in the similes and the narrative responds principally to the poetic need to bridge the simile and the context.

The method used for analyzing the similes is, first, to examine the position of the simile in order to determine at what point in the narrative it occurs. Second, the simile proper is assessed as to length and degree of development, language, internal pictorial content and theme. How does a given simile resemble or differ from the type to which it belongs? Next, it

---

[7] M. Silk, *Interaction in Poetic Imagery* (1974) 16, cites this passage as an example of the classic epic simile. We owe much of our understanding of the interaction of simile and narrative to Fränkel, *Gleichnisse*, cf. *Dichtung und Philosophie des frühen Griechentums*[2] (1962) 44-49 [tr. M. and J. Willis (1975) 40-44], who dispelled the notion argued most recently by G. Jachmann, *Der homerische Schiffskatalog und die Ilias, Arbeitsgemeinsch. für Forsch. des Landes Nordrhein-Westfalen*, Wiss. Abh. 5 (1958) Excurs III, that a single *Vergleichspunkt* must be sought: multiple connections exist between the narrative and developed similes.

[8] Noted by G. S. Kirk, *Homer and the Oral Tradition* (1976) 76f.

is important to weigh the relationship of the simile to its context. At the level of diction the immediate surroundings (which include the resumptive clause and introductory clause) are perused to determine what overlap of vocabulary, if any, occurs with the narrative, or in the case of simile pairs, with the other simile. Which vocabulary is restricted to the simile and which is formulaic? Of special interest are action verbs, epithets, and anthropomorphic descriptions, such as τέκνα for lion whelps (e.g., 17.133), as opposed to σκύμνοι (18.319), which is zoologically correct. Does the explicit animal subject in the simile carry over in the narrative in metaphorical language? How, in short, does the imagery influence the narrative, and how does the narrative influence the simile?

At the level of theme, the simile and its context are analyzed to determine to what extent the action of a simile parallels the action in the narrative. Does an animal subject appear in other contexts in the passage or episode? What, for example, is the relationship between the simile and terms of abuse, scavenging threats, omens, or descriptions of sacrifice? Does the interaction between simile and other contexts involving animal subjects contribute to irony or pathos in the narrative? Does the interaction serve to amplify, foreshadow, or echo events, or does it structure the passage in which it appears? Finally, it is crucial to determine if the simile is part of a larger pattern of imagery in a Book or larger portion of the narrative.[9] In following this method I hope to show beyond a reasonable doubt the unity of the simile at the levels of theme and diction. A close study of the context and the descriptive vocabulary in similes allows us to penetrate further: to look at narrative in which no explicit simile occurs and to discern patterns of subdued animal metaphors. This second approach occasionally proves valuable in the most dramatically charged episodes in which the language tends to have a figurative or secondary meaning. This method again proceeds from the assumption that the integrity of simile and narrative is so fundamental to the epic tradition that the language generates secondary imagery.

The similarity of my approach in certain respects to those of Moulton, Fränkel, Segal, Redfield, and Faust will be obvious.[10] Moulton's study

---

[9] C. Macleod, (above, p. 1, n. 2) 16-28, 43-45, is a good guide to the question of significant repetitions, echoes, foreshadowing, and recurrent patterns in Homer.

[10] Moulton, *Similes*, is the best treatment of the subject in the light of oral theory. Fränkel, *Gleichnisse*; Segal, *Theme;* Redfield, *Nature* 160-223, esp. 193-203; M. Faust,

indicated the direction for a study of the integration of one type of imagery
with narrative, and in concentrating on animal imagery I have found the
connections between simile and narrative to be more specific than Fränkel
suggests, in terms of the implications both for the immediate context and
for entire episodes. Like Segal I see the manipulation of formula, imagery,
and theme as parts of the essential process of oral composition, and I agree
that the repetition of formulae, while always of potential significance, is
often trite or of relative unimportance.[11] Sometimes, of course, the repetition of a word may result from a tendency to use a word again as soon as it
has occurred. And, as the work of Parry and Lord has shown, some correspondences may be imaginary and merely result from the formulaic language in all parts of the poems, including the similes. To complicate matters, our concept of the formula is undergoing changes, as indicated by two
recent studies on the impact of meter on the selection of formulae.[12] Oral
theory does not, however, rule out the possibility of artful concinnity. In
order to distinguish between the artful and the arbitrary the critic must distinguish between isolated instances and significant accumulations of shared
vocabulary. In keeping with current developments in Homeric studies, this
work attempts to demonstrate how the simile, one of the most powerful
poetic devices in the epic, functions at the level of diction and theme, and
how it contributes overall to Homeric artistry.[13]

   This study does not claim to make a typology of all Homeric similes,
but rather limits itself to the principal types involving animals in lion,
herding, and hunting similes in the *Iliad*.[14] Since my approach to the animal
subject lies primarily at the level of theme and diction, zoological aspects of
animals (habitat, weaning practices, etc.) or practices in animal husbandry
are not considered, except in Chapter 2, where the evidence for species in
the poems is compared with documented faunal remains. The interested

---

"Die künstlerische Verwendung von κύων 'Hund' in den homerischen Epen," *Glotta* 48
(1970) 8-31.

[11] Segal, *Theme* 3 and nn. 1 and 2.

[12] T. Jahn, *Zum Wortfeld "Seele-Geist" in der Sprache Homers, Zetemata* 83
(1987), observes a number of words of the same general meaning being used inter-
changeably depending on what metrical positions they have to fill in the line; cf. E.
Visser, *Homerische Versifikationstechnik, Europäische Hochschulschriften*, Reihe 15,
*Kl. Sprachen und Literaturen*, Bd. 34 (1987).

[13] Cf. *Homer: Beyond Oral Poetry. Recent Trends in Homeric Interpretation*, ed. J.
M. Bremer, I. J. F. De Jong, and J. Kalff (1987) vii.

[14] Scott, *Homeric Simile:* Appendix, 190-205.

reader is referred to the appropriate fascicles of *Archaeologia Homerica* and other relevant works in the notes and in the bibliography. The role of the animal in early Greek philosophy, especially as an analogical tool, also strays beyond the province of the present study.[15] The vital role of the animal in Greek religion, especially in sacrifice, is only considered in passing as a simile subject.[16] However, the question of bird epiphanies is relevant, because formal links between simile and epiphany allow for the possibility that beliefs in theriomorphosis inform the metaphorical transformations in the animal similes. Beyond general considerations sketched out below, this work does not assess the position of animals as reflected in the social and economic fabric of the poem. These and other questions of an anthropological nature have been handled well in Schnapp–Gourbeillon's engaging study of the animal in Homer.[17] One could further explore the significance of pastoralism in Homer by making detailed reference to African pastoral societies, but such a comparison does not fall within the scope of the present work. Because of the prevalence of developed animal similes in the *Iliad* and the relative lack of such similes in the *Odyssey* (with notable exceptions), emphasis will be placed on the intersection of animal imagery and narrative in the former. But since the *Odyssey* is so rich in background material about the role of the animal in Homeric society, it will serve as an important source of evidence for recreating attitudes towards animals.

Several reasons have influenced my choice of similes with animal subjects. Animal similes are the most frequent type, and because they lend themselves so well to development, they account for more hexameter verses than any other simile type in the *Iliad*. Because of the frequent occurrence of the animal subject outside of similes they provide for the greatest overlap with the narrative. This overlap is *a priori* meaningful on formal grounds in the case of omens and other contexts which take the form of a simile. Moreover, animal similes, especially the herding and hunting similes, are unified by theme and language. The linguistic unity relies in large part on the strong presence of emotive and psychological vocabulary. This

---

[15] See U. Dierauer, *Tier und Mensch im Denken der Antike, Studien zur antiken Philosophie* 6 (1977), especially "Tiervergleiche und Tiergleichnisse," 6-15; cf. G. Lorenz, *Die Einstellung der Griechen zum Tiere* (diss. Innsbruck 1972).

[16] The standard work on sacrifice is W. Burkert, *Homo Necans* (1972, [tr. P. Bing, 1983]); cf. E. T. Vermeule, *Götterkult, Archaeologia Homerica* V (1974) 95-100.

[17] Schnapp-Gourbeillon, *Lions*.

vocabulary in turn represents the greatest percentage of unrestricted language in the similes. Its concern with heroic ideals implicates the animal simile with the basic human subject of the poem. It must be borne in mind that for the epic audience, far more than for the modern reader, the animal world had a life of its own quite apart from its literary embodiment. Animals, like their human counterparts, are animate and mobile creatures. This is of vital important in a poem which stresses action. Like human beings, animals have faces, eyes, ears, limbs, hearts, digestive systems, and so forth. They are more similar to human beings than, for example, trees or bodies of water, even though these inanimate subjects may share forces with human beings in the Homeric cosmology. The greater similarity and *sympatheia* between people and animals sharpen the portrayal of pathos, which is one of the chief poetic functions of the similes. Unlike storms, rivers, and clouds, animals and human beings have a death as well as a life. This link is crucial for exploring the mortality of the hero.[18]

---

[18] Cf. J. Griffin, *Homer on Life and Death* (1980) *passim.*

# Chapter I
# Animals in the Similes

## The Predominance and Functions of Animal Similes

A study of the interaction of animal imagery and narrative in Homer requires first an examination of the animal simile.[1] Although the epic tradition shows a fondness for images of wings and flight, animal metaphors are relatively rare in the Homeric poems: the developed simile is the vehicle used to suggest the transference of an animal quality to a human subject or inanimate object.[2] Already in antiquity the animal simile was praised for its descriptions of the swift movements of the attacker, the flight or paralysis of the victim, and other purely physical aspects of animals.[3] Psychologically, the animal simile distinguishes itself from other simile groups as a

---

[1] There has been no sustained literary consideration of animal imagery in antiquity or in modern scholarship, with the exception of Edouard Delebecque's comprehensive monograph on the horse (Delebecque, *Cheval*). The study of animal imagery is included in the literature on similes, which has been summarized by Schnapp–Gourbeillon, *Lions* 15–37; cf. Moulton, *Similes* 11–13. Scott, *Homeric Simile* 56–83, briefly treats lion, wolf, deer, hunting, insect, bird, and domestic animal similes. For an excellent recent discussion of similes in general, see M. W. Edwards, *Homer, Poet of the Iliad* (1987) 102–110.

[2] Animal terms of abuse, especially those derived from κύων, are a notable exception; see Chapter 5. Avian metaphors are discussed, along with kennings and metaphors classified by Aristotle as the "animating type," by W. B. Stanford, *Greek Metaphor* (1936) 136; cf. A. Keith, *Simile and Metaphor in Greek Poetry from Homer to Aeschylus* (1914) 14; C. Moulton, "Homeric Metaphor," *CPh* 74 (1979) 279–93. Stanford, *op. cit.* 130, classifies γλαυκῶπις and βοῶπις as "semi–metaphors" whose primitive significance is at least partially forgotten.

[3] Ancient critics particularly admired certain animal similes for their ἀκρίβεια, cf. A. Clausing *Kritik und Exegese der homerischen Gleichnisse im Altertum* (diss. Freiburg 1913), N. J. Richardson, *CQ* 30 (1980) 265–87, and K. Snipes, *AJP* 109 (1988) 196-222. As Eustathius and other ancient commentators pointed out concerning the poetic functions of the simile in general, the device may contribute to the narrative a sense of vividness (ἐνάργεια) or ornament (κόσμος); it may intensify (cf. αὔξεσις) or give relief (cf. διανάπαυσις), and inject pathos. A simile may contribute to the overall οἰκονομία of the poems, for example, in the use of foreshadowing. (Eustathius comments at length on the general function of similes at 176.20ff., 253.24ff., and 1065.29ff.)

medium for exploring the motives and reactions of victor and victim, predator and prey.[4] The tendency to compare human and animal drives through psychological vocabulary lends animal similes a greater degree of unity than for other simile types. The frequent linguistic interaction between animal simile and narrative is symptomatic of an anthropomorphizing tendency that makes lions have "manly spirits" and wasps live in "houses." At the same time a theriomorphic urge can be observed in names and epithets such as Autolykos and θυμολέων, in the divine bird epiphanies, and in ritual animal disguise. The blurring of human and animal attributes in the Homeric poems complements the fascination with hybrids and theriomorphic procreative acts in the Epic cycle and in Greek art and myth. It also betrays the influence of folklore found, for example, in the theme of the self–sacrificing animal parent in omen and simile that would later enter the writings of Physiologus.[5] Closely related to the theme of the animal mother is the paternal relationship of shepherd to flock, as expressed in the military metaphor of the ποιμὴν λαῶν.

Well over half (125:226) of the *Iliad* similes have an animal subject. More than five–sixths of the animal similes occur in battle narrative.[6] The majority of the animal similes in the war epic concern herding and hunting.[7] Unlike the technical or domestic similes, which contrast sharply with the narrative, the herding and hunting similes often reinforce the battle setting. Themes of the predator attacking his prey, or the shepherd guarding his flocks, reproduce in all essentials the relationship between the warrior and his opponent, or between the commander and his troops. Animal subjects

---

[4] The only non–animal simile that possibly anthropomorphizes an inanimate object is found at 14.16ff., where the sea is said to be "watching for" (ὀσσόμενον) the rush of the winds; cf. Leaf, *ad loc.* As B. Simon has shown in his chapter "Mental Life in the Homeric Epics," in *Mind and Madness in Ancient Greece* (1978) 53–77, Homer has a "tendency to ascribe the origins of mental states to forces or agencies outside the person" (p. 62). These external agencies may be divine, but often they are animal forces; cf. U. Dierauer, *Tier und Mensch im Denken der Antike, Studien zur antiken Philosophie* 6 (1977) 8. On pathos in similes see J. Griffin, *CQ* 26 (1976) 162, 180f.

[5] Bird omens: 2.311ff., 12.219ff; similes: 9.323f. (bird), 11.113ff. (doe), 16.259ff. (wasp), 17.4f. (cow), 133ff. (lioness), 18.318ff. (lioness). On parental devotion in Physiologus see, e.g., the pelican, called φιλότεκνος, in D. Offermanns, *Der Physiologus nach den Handschriften G und M* (1966) 29.

[6] Bowra, *Tradition* 123, and others (e.g., Lee, *Similes Compared* 5) explain the higher ratio of *Iliad* to *Odyssey* similes in general by pointing to the concentration of similes, especially developed similes, in the battle books.

[7] Cf. Redfield, *Nature* 189, who counts herding and hunting similes as the largest single group.

are different enough from the human objects of comparison to provide a point of contrast, yet sufficiently alike to transpose human actions and emotions to the animal actors in the similes. The herding and hunting similes involve, if only sometimes by inference, human protagonists, and thus dramatize the basic conflict between man and natural or supernatural forces found elsewhere in river, sea, storm, or fire similes.

Many of the similes which most palpably convey pathos concern the Achaians, who have left behind their families and homelands to regain a stolen bride. Unlike the Trojans, who maintain communal ties, they have lost contact with society. Although digressions and biographies of minor warriors offer glimpses of the domestic lives of the Achaians, the plot furnishes no occasion for the likes of the farewells of Hektor and Andromache, the poignant encounters between Hektor and his son and father, or the laments of Hekabe and Andromache. Many of the great animal similes that mark the strong bonds among the Achaian warriors, such as the mother cow shielding her new–born (17.4ff.), or the lioness tracking the hunter who has stolen her cubs (18.318ff.),[8] counterbalance the domestic encounters possible in the House of Priam. At the same time certain pathetic animal similes serve to correct the Philhellenic bias of the epic by manipulating the sympathy of the audience towards the Trojan victim, who may be compared to the innocent prey in lion and hunting similes.[9] In an example discussed below Agamemnon, even in his own *aristeia*, is made to appear heartless through a simile of a lion tearing out the delicate heart of a pair of fawns while the doe looks on.[10] The emotive function of Homeric animal similes, which Ennius and Virgil later imitated and developed, is prominent in the *Iliad*, where the similes are used as a means of exposing the vulnerability of even the greatest heroes in the face of death.[11]

---

[8] The pathos of this simile was noted by the Townleian scholiast, who commented that nothing could more forcibly (ἐμφαντικώτερον) show Achilles' love for Patroklos than the lioness's reaction to the taking of her young.

[9] The assertion that the poet of the *Iliad* underplayed the defeats of the Achaians in order to glorify their successes at the expense of the Trojans was first made by the bT Scholia. As N. J. Richardson, *CQ* 30 (1980) 274, points out, the allegation is sometimes forced; nevertheless a certain bias, which presumably was inherent in the tradition, is felt. Cf. M. van der Valk, "Homer's Nationalistic Attitude," *AC* 22 (1953) 5–26.

[10] pp. 59–60.

[11] Compare, for example, Virgil's treatment of a lion simile at *Aen.* 10.723ff. with *Iliad* 3.23ff.: *gaudet* (726), cf. ἐχάρη (23). Note the emphasis on emotive states in the wolf simile at *Aen.* 9.59ff.: *ille asper et inprobus ira | saevit ... haud aliter ... ignescunt irae.* For a discussion of the concordance of emotions between man and animal in the

The heightened emotional tone of the simile in general can be appreciated by observing that similes not infrequently appear in charged speeches
when the speaker wishes to exhort or shame a person, or to pity or make
sport of him.[12] A simile may occur where one might expect a term of abuse,
as when Agamemnon in the *Epipolesis* angrily interrogates those holding
back from the fighting, "Why are you simply standing there bewildered,
like young deer I who after they are tired from running through a great
meadow I stand there still, and there is no heart of courage within them?
Thus are you standing still bewildered and are not fighting" (4.243ff.). The
poet may address a simile in a pitying or mocking tone directly to one of
his characters. When Patroklos kills Kebriones, for example, the poet
compares his attack to the leap of a plundering lion who has been wounded
in the chest; but Patroklos is also to die, and to drive home the ominous
point of comparison, the poet directs the resumptive clause to him: ὣς ἐπὶ
Κεβριόνῃ, Πατρόκλεες, ἆλσο μεμαώς (16.754).[13] The Homeric simile
in general is a versatile means for conveying the poet's wide range of
voices.[14]

Animal similes build upon phraseology found in terms of abuse,
omens, epiphanies, scavenging threats, and other traditional contexts involving the animal world; in building upon such contexts, the epic became
progressively shaped and unified in the process of performance. Internal
evidence from both the *Iliad* and *Odyssey* suggests that Homeric audiences
not only enjoyed hearing naturalistic descriptions of animals told for their
own sake, but that they preferred animal over other types of comparison.

---

similes of Virgil and his predecessors, see S. Rocca, *Etiologia virgiliana* (1983) 153.
The potentially manipulative force of similes mentioning the death of a vulnerable or
noble animal is also exploited in Shakespeare's use of animal imagery. In an example
from *Troilus and Cressida* (3.3.161) Ulysses, trying to spur on Achilles by making him
jealous of Ajax, avers that if he does not act he will be "... like a gallant horse fall'n in
first rank, I Lie there for pavement to the abject rear; I O'er run and trampled on." See C.
Spurgeon, *Shakespeare's Imagery* (1935) 27, and A. Yoder, *Animal Analogy in Shakespeare's Character Portrayal* (1947) 47–51.

[12] See Moulton, *Similes* 100, on similes in speech.

[13] Patroklos, upon overcoming Kebriones, delivers a mocking simile (16.745ff.).
On apostrophe see A. Parry, *HSCP* 76 (1972) 9–22; cf. E. Block, "The Narrator Speaks:
Apostrophe in Homer and Virgil," *TAPA* 112 (1982) 7–22.

[14] In a narratological study of the *Iliad* I. J. F. de Jong, *Narrators and Focalizers*
(1987) 135–36, distinguishes between similes occurring in "character–text" (10%) and
"narrator–text" (90%), and shows that while the primary function of similes is to interpret events or persons, those similes assimilated to a character acquire a secondary function of expressing their emotions.

Animals appeal to the familiar and the commonplace, and yet they are veiled in mystery. The animal is at once comprehensible and unknowable.

The frequent representations of animals, not only in similes, but also in sacrifice scenes, omens, folk etymologies, proverbs, parables, terms of abuse, and digressions are idealized mirrors of everyday rural life, speech, and religious practices. The poet is the custodian of this cultural experience.[15] Idealized representations of everyday life form a coherent world of their own that inspired audience belief. The inclusion of animal lore, along with material items such as weaponry and dress helped make the narrative "πιθανός" by providing a realistic backdrop to the extraordinary actions of gods and heroes. The degree of naturalistic portrayal in Homer contrasts with the supernatural, monstrous, or demonic aspects of animals in the Epic Cycle and in many non–Greek epics.[16] In performance, the familiar animal subject responded to an expectation on the part of the audience to hear about matters drawn from the world about them. The use of the familiar had a secondary effect of uniting listeners. There is something ultimately ennobling and comforting about a type of imagery that integrates the audience with the natural world as the epic develops themes of human discord, violence, and death.

## The Oral Nature, Date, and Distribution of Similes

Modern scholarship has shown that similes, like other parts of the poems, were composed according to the techniques of oral poetry. Parataxis, enjambment, and repeated phraseology are in evidence throughout the similes.[17] Similes fall into recurring types. The epic singer had at his command

---

[15] A concise example occurs in Book 12. Hektor shuns Poulydamas' warning about the omen of the eagle and snake: he cares not a whit about the movement of birds, εἴτ' ἐπὶ δεξί' ἴωσι πρὸς ἠῶ τ' ἠέλιόν τε, | εἴτ' ἐπ' ἀριστερὰ τοί γε ποτὶ ζόφον ἠερόεντα (12.239f.). Here we have the epic's version of a formula drawn from the trade of the bird diviner. There is no way of proving, of course, that this is a direct quotation from bird lore, although the epic version in all probability differs in no significant way from actual divinatory formulae in hexameters; cf. *Erga* [ed. West] 801, 828 and note *ad loc.*, and the lost Ὀρνιθομαντεία which West, 46, considers a likely extension of the poem by Hesiod.

[16] Cf. J. Griffin, *JHS* 97 (1977) 39–53.

[17] Two studies, largely overlapping in aim as their shared title, *The Oral Nature of the Homeric Simile* (=J. Hogan [diss. Cornell 1966] and Scott, *Homeric Simile*), suggests, seek to prove the oral composition of the similes. Whereas Hogan concerns him-

a repertoire of similes, or rather types of similes, which he could deploy at his discretion according to the demands of the plot. Certain narrative situations are ripe for, and almost demand, certain types, as Krischer demonstrates in his study of simile types in *aristeiai*.[18] For example, the routing of a large army in battle scenes often suggests a comparison with the lion stampeding herds of cattle. Similes, like other examples of ecphrasis, evoke images of graphic intensity and emotional precision. This heightening effect derives in part from the specialized vocabulary of non–heroic scenes. Equally important is the capacity of similes to develop an internal theme or vignette. In bringing a simile to fruition the singer had a degree of freedom in its phrasing and insertion into the narrative. The characteristic form of the developed simile with an apodosis repeating key words of the protasis readily lent itself to ring–composition, which allowed the poet flexibility in determining the appropriate extent of a simile. The innate form and compositional technique of similes give them the appearance of being an organic whole.

The criteria for judging the formulaic character of similes should be reoriented, given the relatively small quantity of hexameter lines that similes represent, as well as their topical subject matter. If Hainsworth's concept of the formular modification as a repeated word–group (exhibiting flexibility in respect to word–order, separation of constituent words, inser-

---

self primarily with the style and language of the simile itself, Scott does not actually examine the language of similes for enjambment, formulae, etc., but bases his conclusions on the assumption that the poems are oral. Hogan's thorough analysis of the use of enjambment, both necessary and unperiodic, between similes in the *Iliad* and *Odyssey* yields the following statistics: 25% of the similes in the *Odyssey* exhibit enjambment, while only two to three similes in the *Iliad* involving more than two lines (cf. Parry, *MHV* 251–265) have enjambment. 17.742–54 is, however, the most complex period in Homer.

[18] Krischer, *Formale Konventionen* 13–90, distinguishes between two types of *aristeiai*. In the first type, the warrior kills an important hero in a duel, while in the second type the warrior merely proves himself while the duel remains unresolved. Krischer establishes a sequence of conventional elements in *aristeiai*: arming, one–on–one and mass combats, wounding, restoration, duel, and fight over the corpse. (His terminology is given above, p. XII.) Krischer further refines each element into motifs, and argues that specific simile types correspond to each motif. Of the five major *aristeiai* in the *Iliad*, none exhibits every element and motif; in the *aristeiai* of Achilles, Hektor, and Diomedes, Krischer must substitute another conventional motif ("*Ersatzmotiv*") to satisfy the requirements of his scheme. Variations on the motifs, as well as the interchangeability of simile types (e.g., natural phenomena or attacking animals), are as striking as the scheme itself.

tion or omission of particles or prepositions, and inflection[19]) is applied to lion similes, for example, a fairer impression of the formulaic nature of the language emerges than by using Parry's criteria. The density of formulae in lion similes is only slightly higher than for any other type of animal simile. Comparison of the occurrence of formulaic phrases in lion similes and in the narrative of the *Iliad* shows that repeated phrases appear with virtually the same rate of frequency. Nearly the same density of formulae found in lion similes can be observed in all other similes.[20]

The most significant interchange of vocabulary between animal simile and narrative occurs in the use of formulaic language for the emotional and mental states shared by beast and man. It is possible to find in the similes approximately forty phrases emphasizing the heroic ideal of bravery and its opposite, cowardice, e.g., μάλα γὰρ χλωρὸν δέος αἱρεῖ[21] or μέγα φρονέων/–οντε(ς),[22] which reproduce formulae in the narrative. From a linguistic point of view animal similes share more in common with the narrative than other groups, and the point of contact lies in vocabulary commenting on emotions.[23]

Attempts to date the similes have been made in reference to works of art and other archaeological evidence,[24] on linguistic grounds, and by analyzing the structural relationship between short and developed similes. On formal grounds, the evolution of the developed simile from short, end–line

---

[19] Hainsworth, *Flexibility* 36; cf. "Good and Bad Formulae," in *Homer, Tradition and Invention,* ed. B. Fenik (1978) 41–50.

[20] The statistics for formulae in lion similes are detailed in Appendix C.

[21] Simile 17.67; cf. narrative 7.479, 8.77, 14.506 in OCT app.

[22] Simile 11.325, 16.758, 824; cf. narrative 8.553, 13.156, 16.258.

[23] The evidence for emotive vocabulary is given in Appendix B.

[24] I have not found the approaches of Schadewaldt and others who argue for affinities between certain simile subjects and Mycenaean or post–Mycenaean art as convincing as Vermeule's comparisons between the Homeric beast similes and the ornamental vernacular of seventh and sixth century vase painting: W. Schadewaldt, "Die homerische Gleichniswelt und die kretisch–mykenische Kunst," in Ἑρμηνεία. *Festschrift zum 60. Geburtstag O. Regenbogen dargebracht von Schülern und Freunden* (1952) 9–27 (=*Von Homers Welt und Werk³* [1965] 130–154), cf. R. Hampe, *Die Gleichnisse Homers und die Bildkunst seiner Zeit,* (1952); Scott, *Homeric Simile* 166–183; E. T. Vermeule, *Aspects of Death in Early Greek Art and Poetry* (1979) 84–95. There is no evidence that epic poets consciously or unconsciously reflected the subjects of contemporary artists; cf. M. P. Nilsson, *Homer and Mycenae* (1933) 275. As R. Kannicht, "Poetry and Art. Homer and the Monuments Afresh," *CA* 1 (1982) 70–86 and plates 1–6, has shown, artists of the Archaic period tended to produce their own free interpretations of certain general situations described by the epic, such as a shipwreck, as answers of art to poetry.

comparisons of the type ὥστε λέων seems at first glance obvious; only a relative pronoun or conjunction in the next line is necessary to continue the simile *ad lib*. But even if one could point with certainty to places in our text where a short, end–line simile once stood before undergoing development, the number of instances would be relatively small (13x *Il.*, 11x *Od.*). Often ὅτε intervenes between the introductory word of the simile and the named subject, indicating that some development in the next line is already anticipated. The likelihood that a short simile originally stood on its own is greater in the case of traditional types, such as lion or tree; but for less frequent subjects, e.g., beans (13.589), it is unlikely that the simile would have carried much resonance without further development. Many similes appear to be self–contained, integral pieces and not merely "add–ons." The short, traditional similes, such as lion, fire, tree, for which there are analogues in Near Eastern poetry, are, as T. B. L. Webster has shown, among the oldest elements of the Greek epic tradition.[25] But the probable evolution of the developed simile should not be interpreted as a criterion for dating short vs. long similes: the epithet–like short simile in all likelihood persisted alongside the developed simile. Linguistically, similes appear to have reached their present form toward the end of the generative phase, before the poem took shape as a written text, possibly as late as 660 BC.[26] This conclusion accords with the general impression that the similes, as one of the poet's most immediate means of expression, blend with the landscapes and experiences known to Archaic Ionian audiences, much as they merge linguistically with their narrative context.

---

[25] T. B. L. Webster, *From Mycenae to Homer* (1958) 223f.

[26] Cf. W. Burkert, "Das Hunderttorige Theben und die Datierung der *Ilias*," *WS* 10 (1978) 5–21. G. P. Shipp, *Studies in the Language in Homer*[2] (1972) *passim*, shows the high concentration of neologisms in the similes, and in his "General Remarks on Similes" 208–222, argues that similes could be either excised from their context or drastically reduced. (But see the review of P. Chantraine, *RPh* 30 [1953] 73.) The evidence for late forms in similes seems irrefutable in most instances, but the conclusions Shipp implies, i.e., that similes are late additions to the body of the text, are open to discussion. The ratio of new to old forms is not higher in the non–traditional similes (e.g., the technological similes) than in the traditional similes. Shipp's observations are useful for dating the similes as relatively late in the oral tradition, but few would be willing to ascribe them to a post–Homeric phrase of composition, i.e., to accord them an absolutely late date. Assuming that the similes were composed orally like other parts of the poem, the date at which similes entered is relatively unimportant; even if they entered at the latest phase, as the text received written status, they were not simply inserted without regard for their context.

Similes in the *Odyssey* differ from those in the *Iliad* both in language and subject matter as well as in frequency and length.[27] Since the language and subject of the *Odyssey* differ from the war epic, however, this is not a surprising development. Gone, for example, are the honorific similes of predatory animals in *Iliad* fighting scenes, although they resurface in modified form in the Battle of the Suitors (*Od.* 22.402ff., 23.48). Different is the tone of the simile comparing the faithless serving women to birds trapped in snares (*Od.* 22.468ff.), but it is entirely appropriate for their un-heroic deaths. The famous lion simile in the *Nausicaa* (*Od.* 6.130ff.), which Fränkel[28] found so ignoble compared to its *Iliad* counterpart, shows signs of being a masterful parody with erotic overtones of two honorific lion similes in the *Iliad*.[29] In such a finely crafted poem as the *Odyssey* it is reasonable to expect the poet to tailor his similes to the language and subject matter. Indeed this tailor–made quality of the *Odyssey* similes per-haps distinguishes them most fundamentally from the families of similes in the *Iliad*.[30] Proportionately there is a greater variety of subjects in *Odyssey* than *Iliad* similes. The poet appears to rely less on traditional types, but shows a fondness for technological subjects and the family. Some have found the similes of the *Odyssey* more bucolic and sentimental in tone. Again, the tone and setting of the *Odyssey* would encourage such a shift in emphasis. But *Iliad* similes can treat rustic themes in an affective way even in the midst of battle, as in Aias' donkey simile (11.558ff.). Indeed it is this unexpected shift of tone from narrative to simile in the *Iliad* similes that makes them so engaging and effective, and perhaps some of this quality is lost in the *Odyssey*, which lends itself less readily to such dramatic con-

---

[27] Moulton, *Similes* 117–19, discusses the quantitative and qualitative differences; cf. Lee, *Similes Compared* 3–39. Although Lee's statistics and tables, 50–80, are useful, his remarks comparing similes in both poems should be approached with caution. Lee, *op. cit.* 5, assumes that virtually all *Iliad* similes are later than *Odyssey* similes, and explains them as rhapsodic interpolations that relieved audiences grown weary of battle poetry.

[28] Fränkel, *Gleichnisse* 70.

[29] Viz. 5.136ff. and 12.299ff. Compare especially κέλεται δέ ἑ θυμὸς with κέλεται δέ ἑ γαστήρ in the body of the similes (*Od.* 6.133; 12.300) and μίξεσθαι with μίγη in the resumptive clauses (*Od.*6.136; 5.143). Lion similes in the *Odyssey* are discussed by W. T. Magrath, *CJ* 77 (1982) 205–212, and R. Friedrich, *AJP* 102 (1981) 120–137.

[30] Moulton, *Similes* 126–34, shows that the developed similes fall into groups (family and children, lion, bard) with thematic allusions throughout the poems.

trasts.[31] *Odyssey* similes show shifts in content and emphasis. The curious repeated lion–doe simile (*Od.* 4.335ff.=17.126ff.) betrays signs of influence from the fable tradition.[32] The two nightingale similes (*Od.* 19. 518ff., 20.66ff.) are unusual for the amount of explicit mythological material they contain, more so than any *Iliad* simile, including the Pygmy-crane simile (3.3ff.). It has been argued that the *Odyssey* similes display a new interest in exploring the emotions.[33] For example, the tears of Odysseus' companions are compared to calves clamoring about their mothers (*Od.* 10.410ff.), and the cries of joy of the reunited Telemachos and Odysseus are likened to the cries of vultures (*Od.* 16.216ff.). But the psychological function is already pronounced in the *Iliad* animal similes.

Explanations for the more frequent use of animal similes in the *Iliad* are necessarily subjective. R. M. Meyer pointed out long ago that in ancient Norse literature animal similes appear at an earlier stage of the tradition and give way to similes involving human beings.[34] Gonda, an authority on similes in Sanskrit literature, also believes that in oral traditions animal similes appear at an early stage.[35] If one could prove that the *Iliad* significantly predates the *Odyssey*, Gonda's theory might be entertained as a possible explanation for the higher concentration of animal similes in the former. This theory would then receive indirect support from Hogan's statistical study on enjambment, which points to a much more frequent use of unperiodic enjambment in *Odyssey* than in *Iliad* similes, thereby suggesting an earlier date for the latter.[36] But in the absence of compelling proof about the absolute date of the poems,[37] Gonda's hypothesis can only be put forward as a possible suggestion; in the end we are left to surmise that the subject of the *Iliad*, with its underlying theme of human predation, lends itself more readily to the use of animal analogies than that of the *Odyssey*.

---

[31] For more developed contrasts in *Odyssey* similes, see H. P. Foley, "'Reverse Similes' and Sex Roles in the *Odyssey*," *Arethusa* 11 (1978) 14–21.

[32] Fränkel, *Gleichnisse* 70; cf. H. Seyffert, *Die Gleichnisse der Odyssee* (diss. Kiel 1949) 27f. and 28, n. 78, who develops this suggestion.

[33] Seyffert, *op. cit.* 102, attributes the smaller number of similes in the *Odyssey* to a decline in the poetic art of the epic tradition.

[34] R. M. Meyer, *Deutsche Stilistik*[2] (1913) 151.

[35] J. Gonda, *Similes in Sanskrit Literature* (1949) 66; this work is useful for understanding the genesis and function of the simile in general.

[36] J. Hogan *The Oral Nature of the Homeric Simile* (diss. Cornell 1966) Ch. 2.

[37] Cf. R. Janko, *Homer, Hesiod and the Hymns* (1982) 228–31, on relative and absolute dating of early Greek hexameter poetry.

# Chapter II
# Animals in the Poems

## Traditional Themes and Attitudes Towards Animals

The *Iliad* and *Odyssey* are chiefly poetical works but also contain rich cultural information which convey an impression of general attitudes towards animals and of the types of traditional material on which the similes drew.[1] Although the perceptions of animals in the poems reflect eighth, and possibly even seventh, century BC attitudes, the date is relatively unimportant, since sentiments towards animals are slow to change, as evidence from modern Greece suggests.[2] The general picture of the interaction of society and the natural world that emerges from the similes is overwhelmingly realistic and uniform. Occasionally one must make allowance for distortion of naturalism in the similes for poetic effects, for example, the use of pairs of animals for formal reasons to correspond to pairs of characters in the narrative.[3] But overall there is a continuity of attitude exhibited in narrative and simile, for similes do not portray a world which is fundamentally at odds with the background of the poem.

Internal evidence from the early Greek epic paints the picture of an agricultural and pastoral society. The epithet ἀλφηστής in the *Odyssey* indicates that the inhabitants were an agrarian people with a grain–based diet, whereas epithets such as "rich in flocks," the many feasting scenes, and

---

[1] The first book to treat animals in Homer in any systematic manner was O. Körner's *Die homerische Tierwelt*, first published in 1880 and expanded in a second edition in 1930. Körner, a zoologist, maintained that all species mentioned in Homer were frequent along the coast of Asia Minor in Homer's time. On the animal in the Homeric worldview, see H. Rahn, *Studium Generale* 20 (1967) 90–105, cf. *Paideuma* 5 (1953) 277–97, 431–80.

[2] I. Morris, *CA* 5 (1986) 81–138, argues that the Homeric poems are based in the eighth century, not the Dark Ages. The most sustained reconstruction of a society based in the Dark Ages was made by M. I. Finley in *The World of Odysseus* (1954/1979).

[3] E.g., Leonteus and Polypoites are compared to two boars, 12.146ff. R. Conley et al., *European Hog Research Project* (1973) 124, report, "Adult male hogs appear to be solitary individuals. They normally travel alone." Cf. 13.198ff. (the Aiantes compared to a pair of lions).

memorable bucolic characters in the poems speak for the important role of animal husbandry.[4] Cattle formed the basis of the economic system: social prestige in as fundamental an institution as marriage was measured by live-stock holdings. Domesticated animals provided food, clothing, and various by–products; teams of sweating oxen reduced the farmer's labor in the fields, and the dog kept off human and animal robbers. Many shared, like Eumaios, a bed in common with their domesticated animals. Beyond the cultivated areas wild animals—rabbit, deer, and the boar—provided sport for the huntsman and his hounds and trained adolescents for war. On the peripheries lived the cattle–robbers and scavengers, the lion and jackal (sc. θώς), who preyed on men's cattle and ease of mind.

Although the epic tradition presents an idealized mirror of society in poems which are to a large extent fictional, it is reasonable to infer that, as in any pastoral society, contact with animals was widespread and intimate. Hunting and fishing, animal sacrifice, divination through bird signs, the care and protection of flocks from predators, or the raising of pigs, were part of daily life. The appeal of such subjects in similes lies in the ready identification possible for the audience and the heroicized stature these pastoral activities attain when juxtaposed with the exploits of heroes whose κλέα the bard immortalizes in song.

The appearance, on all levels of the epic, of the shepherd–flock relationship shows how deeply ingrained it had become in the epic tradition. Metaphorically, the chief commander is known as the "shepherd of the people." In the background of the poems we learn that Apollo had once tended Laomedon's cattle (21.448f.); warriors like Melanippos herded cattle as their occupation before taking up the Trojan cause (15.547f.); even Polyphemos is cast as a shepherd (*Od.* 9.217, etc.). It is this kind of familiarity with pastoral life that makes it natural for Homer to portray Proteus as a shepherd counting and dividing his seals (accorded a cattle epithet, ζατρεφής) and bedding down among them, as the short simile says, "like a shepherd among his sheep" (*Od.* 4.413).

Just as Xenophon shows that hunters are warriors, and warriors hunters (Xen., *Cyn.* 1.1ff.), so the shepherd in times of war is a soldier, and the chief commander is metaphorically the protector of his flock of men. In

---

[4] K. Vickery, *Food in Ancient Greece, Illinois Studies in the Social Sciences* 20 (1936) 48ff.; Richter, *Landwirtschaft* 87ff.

the lion vignette on the Shield the shepherds "march," ἐστιχόωντο,[5] and the predator attacks the bull in the "foremost ranks" of cattle, ἐν πρώτῃσι βόεσσι.[6] The proximity of the metaphorical and literal uses of the shepherd of the host and the shepherd of flocks emerges most clearly in a simile at 13.492ff., comparing Aineias to a herdsman. In this passage ποιμήν occurs in 493, λαῶν in 495.[7] Aineias prepares to lead a mass attack and calls to his side the Trojan lords, Paris, Deiphobos, and Agenor (rather like a shepherd calling the lead sheep by name), and the people trail behind them "as when the sheep follow the lead-ram I as they leave the pasture to drink, and make proud the heart of the shepherd, I and thus also the heart of Aineias was gladdened within him I as he saw the swarm of the host following his own leadership." There are two leaders in the simile, the ram and the shepherd, and Aineias is comparable to both. The qualities of leadership, as well as the massed configuration of the flocks, make the shepherd–flock metaphor appropriate for the commander and his troops. The aptness of the analogy also carries over into the spiritual realm, as emphasized here by the point of comparison which is an emotive one: the shepherd is inwardly pleased in his heart because of the orderliness of his flocks, γάνυται δ' ἄρα τε φρένα ποιμήν (493), just as Aineias rejoices in his breast, ἐνὶ στήθεσσι γεγήθει (494) at the obedient response of the army.[8]

---

[5] 18.577, cf. 2.92 in a martial context.

[6] 18.579 (or simile 15.634), cf. 4.341, 5.536, 8.337, 11.61, etc.

[7] Stanford, *Greek Metaphor* (1936) 135, uses the epithet and its equivalent in the simile at 2.474ff. to demonstrate the similarities and differences between simile and metaphor; cf. C. Moulton, *CPh* 74 (1979) 279f., on the interaction of simile and metaphor. ποιμήν, the substantive properly denoting herdsman (either of cattle or sheep) appears more often in its metaphorical sense as an honorific title than in its literal sense. On the Achaian side ποιμένι or –α λαῶν occurs most frequently with the name of the commander-in-chief, Agamemnon (ten times in the *Iliad*; three times in the *Odyssey*). On the Trojan side, Aineias is called "shepherd of the people" three times in Book 5. Hektor, who has his own line-end epithet (χαλκοκορυστῇ/ήν), is not accorded the title as freely as Agamemnon; he is called "shepherd of the people" only four times (10.406, 14.423, 15.262, 22.277).

[8] This sort of juxtaposition of narrative and simile occurs again with an alternative word for herdsman, βουκόλος. At 15.547f. we hear of the cow–herding activities of Melanippos prior to the war. In the simile (586ff.), a lion runs away because he has done something evil: he has killed either a dog or a βουκόλος. Since this is the only time the death of a man is mentioned in a simile, it certainly would be odd if it did not refer to Melanippos, who is killed. By including the word in the simile the peaceable herding function of the man is recalled and placed in contrast to the fate of the soldier.

Hunting as portrayed in Homer is not so much a necessary activity as a pursuit of adventure.[9] Only once is an animal explicitly procured for food: Odysseus slays a huge stag sent by Circe as a lure and then divides the kill among his companions (*Od.* 10.156–84). The lion appears only incidentally in hunting similes, and never actually as the quarry of the hunter in the forest (except when presented as an alternative to the boar).[10] The leopard is the object of the hunt in a developed simile once (21. 573ff.). Descriptions of the hunt in simile and narrative remind us that superhuman forces control the realm of the forest as they control the affairs of men in society and on the battlefield. Circe dispatches a stag, and Artemis unleashes a boar, and in an enigmatic simile that reflects the theme of divine intervention on the battlefield, an anonymous spirit sends a lion to chase off jackals scavenging on the carcass of a stricken deer (11.480). Like prowess in war, skill in the hunt is divinely inspired. As a kind of epitaph for Skamandrios, one of the victims in Diomedes' *aristeia*, it is said that he was knowledgeable in the hunt, αἵμονα θήρης, and a good hunter, ἐσθλὸν θηρητῆρα, thanks to Artemis; but the goddess of the hunt can be of no use to him now (5.50f.).

Attempts to reconstruct attitudes towards animals from the internal evidence of the poems alone run the risk of becoming circular. For a corroboration of attitudes conveyed by the Homeric portrayal we may turn to modern ethnographic studies of sheep–herding villages, since the pastoral scene in certain pockets of Greece and the Aegean has remained static.[11] People live in close proximity to domestic animals, but they are not kept as pets, nor are they given names, except when the shepherd finds it useful to

---

[9] For a survey of Homeric passages relating to hunting and fishing, as well as the archaeological and artistic evidence for these pursuits from the Bronze Age through the 7th c., see H.-G. Buchholz et al., *Jagd und Fischfang, Archaeologia Homerica* J (1973). For a summary of the scholiasts' remarks on fishing, see M. Schmidt, *Die Erklärungen zum Weltbild Homers und zur Kultur der Heroenzeit in den bT – Scholien zur Ilias* (1976) 182–187, who also summarizes the debate from Plato onwards whether or not Homeric heroes ate fish. For hunting terms in Homer, especially θηρ- and ἀγ-, see the important article by P. Chantraine, *Études et Commentaires* 24 (1956) 31–96; J. K. Anderson, *Hunting in the Ancient World* (1985) 1–16.

[10] A lion drives off scavengers feeding on the carcass of a stag wounded by a hunter's arrow (11.474ff.). A lioness discovers that a hunter has stolen her cubs (18.318ff.).

[11] Cf. R. and E. Blum, *The Dangerous Hour. The Lore of Crisis and Mystery in Rural Greece* (1970) 267, who compare three villages in modern Greece with a "Homeric–period" village. For a survey of the ancient Greek evidence see S. H. Lonsdale, *G&R* 26 (1979) 146–159.

keep track of the flock by designating the lead sheep.[12] A study of the nomadic Sarakatsani of Zagori indicates an almost symbiotic relationship between herdsman and animal; despite this solidarity the shepherd is neither gentle nor sentimental towards his flock.[13] In the Homeric poems the only animals to receive appellations are certain privileged horses and the hound Argos; we never hear of names for cows or any other domestic animals, although we know this to have been a Bronze Age practice from a group of Knossos Ch tablets.[14] Polyphemos addresses his lead ram only as κριὲ πέπον (*Od.* 9.447). Although stems of words for animals frequently give rise to proper names in the epic, those of human beings are not used for animals.[15] Horses are given names that characterize the creature's speed or the color of its coat.[16] Argos likewise denotes a color.[17] In modern Greece dogs guard the flocks and perform salutary scavenging services by

---

[12] Cf. J. du Boulay, *Portrait of a Greek Mountain Village* (1970) 15.

[13] J. K. Campbell, *Honour, Family and Patronage. A Study of the Institutions and Moral Values in a Greek Mountain Community* (1964) 19–30.

[14] J. Chadwick and M. Ventris, *Documents in Mycenaean Greek*[2] (1973) 105.

[15] ἱππ- stems give rise to two dozen proper names and seventeen epithets. Delebecque, *Cheval* 41f., shows a tendency to accord horse–taming epithets collectively to the Trojans. The stem λυκ- occurs in nine proper names, including Autolykos ("Self–Wolf"). λέων/λεοντ- occurs in the name of the Lapith Leonteus (12.188), a warrior of the older generation, as the stem in -ευς indicates. He earned fame, and probably his honorific leonine name, because he fought the shaggy race of centaurs; when he defends the Achaian gates with Polypoietes, he and his comrade are compared to boars (12. 146ff.). The centaurs, called Κένταυροι and Φῆρες, are mentioned in the Catalogue (2.743); cf. Cheiron as healer (4.219, 11.832) and as donor of the ashen spear (16.143, 19.390). About Demoleon, we know only that he was the son of Antenor, good at warding off fighting (ἐσθλὸς ἀλεξητὴρ μάχης), at least until slain by Achilles. His name "Common Lion" may distinguish him from the sovereign lion to whom Achilles has earlier been compared in the monumental simile of a lion whom the entire community, πᾶς δῆμος, has turned out to kill (20.396, cf. 20.164ff.). As an epithet λέων occurs in the compound θυμολέων. Four warriors merit the comparison: Tydeus, Herakles, Achilles, and Odysseus in the *Odyssey*. The βο– stem occurs in the name of the bastard son of Laomedon, Boukolion (6.22), cf. βουκόλος, and is used in feminine names for marriageable girls to attract dowry: Periboia (21.142), and παρθένοι ἀλφεσίβοιαι on the shield of Achilles (18.593). βοῶπις is not reserved for goddesses: one of Helen's handmaids at Troy is so called (3.144). In the Meleager paradigm Kleopatra is said to be called Alkyone (9.562ff.) because her mother mourned her husband with the sorrowful voice of the halcyon: the female was thought to utter a continuous cry when separated from the male.

[16] Delebecque, *Cheval* 146.

[17] In historical times the Greeks named dogs, but did not give human names either to hunting dogs or to pet dogs. Xenophon (*Cyn.* 7.5) provides as examples Thymos, Chara, Psyche, or Hybris.

disposing of the carcasses of horses and other large animals near the village.[18] In Homer, however, the useful herding activity of dogs contrasts sharply with the threat of the scavenger.

Farm animals are considered a part of the community, but they occupy the lowest position in the social hierarchy. Attitudes towards individual domestic animals depend on the relative benefit of a given species to the community. A sheepdog, for example, will earn the affection of the shepherd through its cooperation in moving and protecting the flocks. A mule will be patiently tolerated because it can be coaxed to carry burdens to and from the village. When an animal becomes a nuisance it will not be indulged. Here we are reminded of Aias' simile of the boys trying to chase the hungry donkey out of the cornfield (11.558ff.). The overriding appreciation for work animals in Homer is captured in the simile comparing the two Aiantes to oxen sweating under the plough (13.703ff.). But individual farm animals are not the object of special affection. This impersonal regard for domestic animals as anonymous members of a larger group correlates with the Homeric portrayal of herds of cattle. Except when the lead ram (3.196ff.) or the bull (2.480f.) is singled out, they remain unremarkable.[19] Apart from the horse, it is the wild animals preying on the herds who are admired for particular qualities resembling heroic ideals. The often pejorative attitude towards herd animals in the similes does not reflect an inherent disdain for them; rather it indicates how, through similes, the poet can manipulate the feelings of his audience to side with the individual hero, or shift their sympathies from the victorious hero to the victim.

Saunders, in his work on the people of rural Greece, stresses their intimacy with nature, tempered with respect for its mystery.[20] The Greek farmer will study nature for signs, especially animal portents, and react accordingly. The sighting of two snakes fighting and devouring one another, for example, heralds the death of a close relative. A crow passing on the right augurs well, whereas one flying left augurs ill. Although Christian attitudes have shaped some of these beliefs, the same tendency to interpret and react to phenomena in nature can be seen in the omens and dreams of

---

[18] E. Friedl, *Vasilika. A Village in Modern Greece* (1962) 15f.

[19] Schnapp–Gourbeillon, *Lions* 28ff., shows that the Homeric portrayal of animals in groups often tends to be negative.

[20] I. Saunders, *The Rainbow in the Rock* (1962) 15f.

the *Iliad* and *Odyssey*.[21] Hektor's disdain for bird signs (12.239f.) blinds him to their import in predicting his peril. Despite the intimacy which the Greek farmer feels with the natural world, he is still at conflict with nature. A study of a sheep–herding village in the Pindus valley characterizes the relationship of the peasant in Greece to nature as an *agon*.[22] In the majority of the hunting and herding similes in Homer nature is portrayed as harsh and menacing. This agonistic spirit is captured by the simile of the goatherd who, upon seeing an impending storm, seeks refuge with his flocks in a cave (4.275ff.).

## Animal Species: Principles of Selection

Sixty–two identifiable species of animals are included in the Homeric poems.[23] Apart from the horse, boar, and cattle—and in the *Odyssey* dog and birds—all are restricted to the similes, without which we would have a very incomplete view of the Homeric animal world. Most of the animals referred to in the poems are still indigenous to the region, and for certain extinct species, as we will see, bone samples have turned up to confirm their existence in antiquity. Mammals, birds, fish, insects, and reptiles are represented in the poems; there are few monsters or mythical beasts, with notable exceptions. It has been noted that there is no generic word for animals in Homer.[24] θήρ is usually a synonym for λέων, and the plural indicates a wild land–creature (as opposed to sea–creatures or birds, *Od.* 24. 292, cf. *Erga* 277) or the centaur, sc. φήρ (1.268, 2.743). The diminutive θηρίον designates the stag sent by Circe. τὰ βοτά, meaning "grazing animals," occurs once in the description of the shield of Achilles (18.521). The lack of a common term for fauna in Homer does not so much indicate a failure to distinguish between human and animal (generic terms are rare in

---

[21] Cf. N. Austin, *Arion* 1/2 (1973–4) 219ff., who gives examples of reading natural signs.

[22] Friedl, *op. cit.* 75.

[23] See Appendix A for a list of identifiable species and names in Greek.

[24] H. Rahn, *Studium Generale* 20 (1967) 92f., cf. U. Dierauer, *Tier und Mensch im Denken der Antike, Studien zur antiken Philosophie* 6 (1977) 6f. θήρ is used to distinguish land animals from fish and birds in Hesiod, *Erga* 277. τὸ ζῷον, the generic word for animals in Greek, does not appear until the 5th century (Hdt. 5.10).

Homer), as a tendency to react to each individual species with distinct attitudes. It was not until later that the Greek philosophical tradition established common denominators for all animal species and compared the relative virtues of animals and men.[25]

Among domestic animals, cattle, pigs, sheep, goats, and dogs are frequent, but horses—for the most part restricted to the *Iliad* where they are
constant actors—are the most ubiquitous domesticated animal in Homer.
By contrast with the prevalent lion the leopard is rarely mentioned. The
lion's victims in the forest are the deer and the wild goat; the wolf and his
scavenging cousin the jackal feast on their carcasses. The ferocious boar
offers the huntsman thrilling sport.

Numerous land and sea–birds inhabit the poems. The hawk and eagle
attack smaller birds, such as doves, sparrows, and daws. The vulture is the
airborne counterpart of the "pariah" dog. Other avian species include the
nightingale, the swallow, halcyon, and the heron whose cry cheers Odysseus and Diomedes as they set out on their nocturnal scouting expedition
(10.274ff.). Geese, cranes, and long–necked swans are migratory birds.
In a simile at 2.459ff. (cf. 15.690ff.) they alight along the streams of the
Kaÿstros, one of the few place names mentioned in the similes which
suggests an Ionian background. Among sea–birds, the tern skims the surface of the wine–dark sea (5.51), and sea–crows bob up and down with
the tide (*Od.* 12.418=14.308).

Swarming insects in Homer include flies, bees, and wasps, and the
οἶστρος that maddens cattle. Choruses of cicadas are mentioned, and locusts huddle in a river to escape a blazing fire (21.12–14). The scorpion
and spider are conspicuously absent, as are lice.[26] The water–snake and the
mountain–snake are the only reptiles. No individual species of fish are
named; ἰχθύς serves for all. Identifiable sea mammals are limited to the
dolphin and seal. κῆτος is a general term for sea–monsters, e.g., *Od.* 12.
97. The seadog, κύων, has been tentatively identified as the shark.[27] The
worm and the slithering eel are corpse–eaters. The octopus is a cephalopod
to which the ship–wrecked Odysseus is compared (*Od.* 5.432ff.).

The catalogue of identifiable species suggests at first glance a somewhat random principle of selection. Animals are included for their intrinsic

---

[25] U. Dierauer, *op. cit.* 6f.

[26] Cf. M. Davies and J. Kathirithamby, *Greek Insects* (Oxford 1986) 7–10.

[27] *KlPauly II* (1967) 918f. s. *Hai.*

beauty, energy, and lyrical qualities—qualities that make them seem at times almost supernatural. But to a large degree the themes of the epic influenced the choice. The formulaic phrase for geese, cranes, and long-throated swans, for example, occurs in similes that emphasize the din of battle. The striking number of noxious creatures owes itself to the pervasive scavenging theme that reflects cultural attitudes about an improper burial.[28] The hero's essential quality of fearlessness gives rise to comparisons with animals who will not shrink from attacking men. These are the lion, boar, snake, and leopard, the four creatures into which Proteus transforms himself (*Od.* 4.456f.). The relationship between victor and victim in the war epic dictates pairs of species which are natural enemies in the air, on land, and in water. The lion and cattle of the similes give way appropriately to dolphin and small fish when the fighting shifts to the River in Book 21. The horse, dog, and bird are included in part because they correspond to the speed that distinguishes the successful warrior from the straggler.

The evidence furnished by the systematic study of faunal remains from archaeological sites in the Aegean corroborates the Homeric evidence for familiar species.[29] With the exception of the lion,[30] all animals, both wild and domestic, which frequently appear in Homer have been found at Nichoria in Messenia: cow, sheep, goat, deer, pig, horse, wild boar, domestic and wild dogs, mules, and red and roe deer. Goats and sheep were the most numerous animal population, followed by pigs and wild boar; but goats were twice as prevalent as sheep. Cows, surprisingly, accounted for 63% (by weight) of animal protein in "DA I–III" (1050–775 BC), a dra-

---

[28] Cf. Rahn *op. cit.*, 101f.

[29] Osteological investigation is a relatively new feature of archaeology, first carried out in a comprehensive way in Aegean archaeology by Nils–G. Gejvall of Professor Caskey's team during the excavations at Lerna (J. Caskey, ed. *Lerna:* Vol. I *The Fauna*, by Nils–G. Gejvall [1969]), although Carl Blegen's publication on Troy contains information about the animal remains (C. W. Blegen et al., *Troy* [1958] I, 10 and *passim*; cf. II, *passim*). More recently R. Sloan and M. Duncan have published the faunal remains at Nichoria in Messenia in G. Rapp and S. Aschenbrenner, eds., *Excavations at Nichoria in Southwest Greece* (1978) I 60–76. The general picture that emerges for Lerna and Nichoria is similar enough to take Nichoria as typical of dietary customs and practices in animal husbandry and hunting for the LH and DA. For general information on faunal remains in archaeology, see R. Chaplain, *The Study of Animal Bones from Archaeological Sites* (1971); for mammals of the Caucasus, see N. Vereskshagin, *The Mammals of the Caucasus* (1967). For sacrificial remains in the Greek world, see M. Andronikos, *Totenkult, Archaeologia Homerica* W (1968), and J. Boardman and D. Kurtz, *Greek Burial Practices* (1974).

[30] The bone evidence for lions is summarized on p. 104, n. 3.

matic increase in beef consumption compared with LH dietary customs.
There was a major shift from dairy practices to meat ranching at the transi-
tion from LH to DA. Although horse meat was not eaten, the dog was a
source of animal protein in MH. The population of *canis familiaris* was
moderately large, and about 10% greater than that of the wild dog, or jackal. The canine population swelled in the late DA as hunting increased. The
greatest pressure was exerted on red deer in the DA, and the species is now
extinct. The wild boar was locally exterminated by LHIIIB. The major dietary change between the Bronze Age and the Dark Age was from milk and
cheese to meat as the major source of protein. Whereas sheep, goat, and
cows were mainly raised for dairy production during the Bronze Age, they
were later used with increasing frequency to provide red meat in the diet.

One is tempted to see the portrayal in the Homeric poems as something
of a composite of BA and late DA practices. While dairy production is
vividly evoked, for example, at 2.471 in the chain of similes preceding the
Catalogue of Ships, a primarily meat diet is clearly suggested in the feasting scenes, with the choice viands offered to warriors like Sarpedon (12.
311) or to Aias after the duel with Hektor (7.200ff.). Athenaios (1.9) cites
this last example among other Homeric passages in support of the ancient
view of the heroic image: heroes confine themselves to a strict diet of roast
beef, scrupulously shunning fish and delicacies such as stuffed grape
leaves.


### Names, Gender, and Epithets for Animals

Homeric language for animals is characterized by a concern for naturalism. An extensive system of epithets and formulaic phrases points to
prominent physical aspects, movements, habits, and habitats of species.
The range and detail of description reflect close observation of nature.[31] In
addition to naturalistic portrayal, a second system of epithets and formulae
characterizes the inner states and qualities of frequently recurring species
such as the lion, boar, dog, and horse. In a sense these and other animals
are interpreted by the oral tradition as symbols, but without distorting a

---

[31] Among modern critics impressed by the soundness of Homeric zoology and
naturalistic portrayal, see in addition to Körner, *Homerische Tierwelt*, H. W. Auden, *CR*
10 (1896) 107, L. Meule, *Soc. de Zoo. de Paris* (1910), J. Borasten, *JHS* 31 (1911)
216–50, M. Schuster, *MVPhW* (1928) 28–46, A. Severyns, *BCH* 70 (1946) 540–47, J.
Duchemin, *REG* 73 (1960) 362–415; *contra*: Schnapp–Gourbeillon, *Lions* 12.

fundamentally naturalistic depiction. Because the symbolic epithets and formulae insist on shared spiritual forces in human beings and animals, they reinforce the basic intimacy with the animal kingdom suggested by the first system. Thus, the representation of animals in Homer is concerned with both naturalism and *concordia*. The composite poetic portrait of the animal reflects and exploits the ambiguity inherent in human perceptions of animals.

An examination of species names in terms of the number of syllables, their gender, and epithets reveals attitudes about animals and about how the epic tradition manipulates traditional material to poetic ends. Species names for animals in the nominative range between one and four syllables. While most species are disyllabic, there is a tendency for domesticated creatures living close to people to be one and two syllables, for example αἴξ, βοῦς, ὄις, σῦς. On the other hand wild species, especially birds, which live at some remove from civilization tend to be trisyllabic, for example αἰγυπιός, αἰετός, ἔλαφος, λαγώος, πάρδαλις, or κνώδαλον for any wild or dangerous animal. θήρ is an exception to this observation. There are no domesticated species of more than two syllables, except when an epithet reflecting practices in animal husbandry or sacrifice is added.[32] There are monosyllabic nouns for wild creatures, but these rarely appear alone and are usually spun out into several syllables by the use of epithets emphasizing their wild habits or habitats, e.g. ὠμοφάγοι θῶες. The practice of making a wild animal more imposing by expanding its name can be seen in the description of the destructive boar sent by Artemis to ravage Oineus' crops. Normally the boar is distinguished from the domestic pig by adding ἄγριος, κάπριος (cf. κάπρος), or ἀγρότερος, but the Kalydonian boar is depicted in the alliterative phrase not only as wild but as having white tusks: σῦν ἄγριον ἀργιόδοντα (9.539).

Most species names for animals are masculine, although a few, including πάρδαλις and ἄρκτος, are feminine. But several species admit a distinction between masculine or feminine. In the similes and metaphorical uses of these double–gender species the ambiguity is sometimes exploited. The meaning then becomes pejorative or pathetic. When Agamemnon rebukes his men for not fighting he asks, "why do you stand there stunned like fawns?" As the comparison develops he uses the feminine relative pronoun αἱ at the beginning of the next verse. Since νεβρός, the word for

---

[32] For example, σίαλος for full–grown, domestic swine, is properly an epithet joined with σῦς.

fawn, can be either masculine or feminine, the use of the feminine here effectively increases the criticism implied by Agamemnon that his men are behaving in a womanish manner. Still, there is no distortion of naturalism for poetic effect. Here we may observe an overlapping of the literal meaning of "fawns" in the feminine with its metaphorical meaning.

The use of the feminine gender sometimes adds pathos. In Achilles' *aristeia* Agenor is compared to a leopard who, though pierced (feminine participle πεπαρμένη, 21.577) by a hunting spear, still does not lose her fighting spirit. The feminine gender here does not emasculate Agenor but rather sharpens our sympathy for Achilles' last victim before his encounter with Hektor. The pathetic coloring is pronounced in similes portraying the relationship between mother animal and offspring, as in the simile comparing Menelaos astride the corpse of Patroklos to a cow over her first–born, or Achilles' simile of the lioness bereft of her young.[33]

One is impressed in Homer by the generous use of epithets and short descriptive phrases for animals. Virtually every species mentioned, both wild and domestic, has at least one epithet, and a large majority can boast between three and five. From the point of view of formulaic composition, economy is sometimes suspended. A frequently occurring species, the eagle, has six superlatives, four of them appearing at end–line position with πετεηνῶν.[34] The extravagance of animal epithets is not so much an indication of subtle distinction between various species of animals which cannot be apprehended by the modern reader (although certain epithets, like χλούνης of the boar, have baffled commentators for many centuries), as a

---

[33] Λέων or λίς are used for both male and female. λέαινα is first attested in A., *Ag.* 1258.

[34] τελειότατον (8.247=24.315) and ἐλαφρότατος (22.139=*Od.* 13.87), though different in meaning, are metrically interchangeable (with use of ν-moveable in the word preceding ἐλαφρότατος); cf. ὤκιστος (15.238, 21.253). The eagle appears sixteen times and has an impressive number of epithets totaling eighteen. The interest in the eagle no doubt stems from the bird's connections with Zeus and its important role in bird–divination. As Zeus' bird (8.251) many of the eagle's epithets are superlatives: φίλτατος suggests his intimate link with Zeus; the eagle is the strongest and largest of birds κάρτιστος, μέγιστος, and like a good messenger (21.253) is the swiftest: ὤκιστος (cf. μετὰ πνοιῆς ἀνέμοιο *Od.* 2.148). The eagle flies high above the earth (ὑψιπέτης) and is renowned for his sharp–sighted vision: ὀξύτατον δέρκεσθαι ὑπουρανίων πετεηνῶν. In omens the eagle is the one who brings the surest presage of fulfilment, τελείοτατος, and bodes well by flying to the right. ἀγκυλοχείλης refers to the vicious talons, and ὁ θηρητήρ gives an index of the eagle's behavior. At least two varieties are involved: μόρφνος (24.316, cf. μέλας, 21.252) denotes the swamp–eagle, and a periphrasis, ὅν καὶ περκνὸν καλέουσιν, a dappled variety. See D.'A. W. Thompson, *A Glossary of Greek Birds*[2] (1936) 255.

reminder that rival poets compete with audiences by displaying knowledge of animal, especially horse, lore. Judging from the high frequency with which βοῦς and its derivatives and relatives occur, the cow was more familiar to the epic audience than any other domesticated animal;[35] this impression is corroborated by the osteological evidence and suggests that the portrayal of this valuable livestock animal was not simply glamorized to suit the heroic tenor of the epic. The rich and highly–developed system of epithets for animals in general contrasts with the sparse description of other non–human objects—the flora, the physical setting, and meteorological phenomena.

Epithets depict the habitat of animals,[36] their habits or characteristics,[37] and single out salient visual details, such as the mane of the lion, ἠυγένειος, or the wing–span of birds τανυσίπτερος, τανυπτέρυξ. Typically the habitat in similes ranges from the country to the mountains, which occasionally blend with the locale of Mount Ida in the *Iliad*.[38] The relationship between predator and prey found in the internal action of so many similes is telescoped into an epithet like φασσοφόνος (15.238) for the hunting habits of the murderous hawk, likened to Achilles making a swoop upon Hektor (22.139ff.). Sometimes the epithet will point up a feature related to a practical function the animal serves, e. g., εἰροπόκος for the wool of sheep, or an outstandingly beautiful feature: ἐριαύχην for the horse's elegantly arched neck. Color epithets are frequent but rarely go beyond the fundamental distinction in Homer between light and dark.[39] Often, of course, an epithet may be explained by metrical requirements. An

---

[35] Next to the horse, the cow and bull have the highest number of epithets; cf. Richter, *Landwirtschaft* 46–49. The cow was probably introduced from the northerly parts of Western Asia early on in the Bronze Age; G. Zeuner, *A History of the Domestication of Animals* (1963) 214ff.). The words for bull, cow, and their offspring, πόρτις belong to the Indo–European family; E. Boisacq, *Dictionnaire étymologique de la langue grecque* (1923) 804, s. πόρ[τ]ις.

[36] E. g., ὀρέστερος serves for the snake, ὀρεσίτροφος for the lion, and ὄρεσφι(ν) for the jackal and eagle, ἄγραυλος for oxen, ἁλιοτρεφής for seal, εἰνάλιος for sea cows.

[37] E. g., ὠμοφάγος for the jackal's predilection for flesh, χαμαιευνάς for the pig's prostrate life–style.

[38] E. g., Ἴδης (11.105; 111: narrative), cf. ὕλην (11.118: lion simile).

[39] Thus αἴθων modifies cattle, horses, lion, and the eagle; ἀργός (cf. ἀργεννός) cattle, dogs, sheep; μέλας is used for the eagle, and for lambs and sheep destined for chthonic sacrifice. The wolf is of a lightish hue, probably grey, πολιός. φοῖνιξ describes the color of the bay–horse. See C. J. Rowe, "Conceptions of Colour and Colour Symbolism in the Ancient World," *Eranos Jahrbuch* 41 (1972) 327–64.

accumulation of epithets can lend a note of pathos, as when Andromache recalls before Hektor the death of her seven brothers while they were "tending their white sheep and their lumbering oxen" (6.424).

If one contrasts the Homeric flora with the fauna, the more than fifty species of trees, plants and shrubs is somewhat less than that of zoological species. But the ratio of botanical similes to animal similes is 2:15. The *Odyssey* is remarkable for a few elaborate descriptions of plant–life in *loca amoena* such as Alkinoos' garden (*Od.* 7.114ff.), or Kalypso's cave (*Od.* 5.63ff.). In the recognition scene between Odysseus and Laertes Odysseus identifies the ten apple–trees, thirteen pear–trees, forty fig–trees and fifty rows of vines which his father had designated to him as a child (*Od.* 24. 336ff.). Apart from this delightful passage and *loca amoena*, the description of plant–life is not followed for its own sake. Individual plant names, such as lotus and the mysterious moly, are mentioned in passing in Odysseus' travels. In the *Iliad* names of trees and plants occur almost exclusively as simile subjects. The minor warrior Gorgythion has been immortalized by the memorable simile comparing his severed head to a garden poppy drooping under the weight of spring rains (8.306f.); Thetis raised Achilles like a sapling in an orchard, and he grew up like a shoot (18.56f.; 437f.). There are relatively few epithets for plants and trees: even in the most elaborate botanical description, the garden of Alkinoos, there are only two epithets in eighteen lines mentioning six species: ἀγλαόκαρποι of apples (*Od.* 7.115), and γλυκεραί of figs (116).[40]

As we would expect in an oral poem, onomatopoeic words are common and show a preference for animal sounds.[41] μυκηθμοῦ ... βοῶν (*Od.* 12.265, 10.413, 12.395), for example, reproduces the lowing of cattle. From the lion vignette on the Shield we have a literal usage of the verb for the groaning of the lion's victims (μεμυκώς, 18.580). Figuratively, μυκάομαι describes the groaning noise of the portals of Olympos (5.749, 8.393, cf. 12.460); the doors of Odysseus' palace groan "like a bull" (*Od.* 21.48). The  sound is then sometimes transferred to an inanimate  object with an animating effect. Another figurative use of μυκάομαι from Book 20

---

[40] E.g., ὑψηλός refers to the loftiness of a tree, such as the silver fir, ἐλάτῃσιν ἐοικότες ὑψηλῇσι (5.560). μελιηδής modifies the dog's–tooth grass (ἄγρωστις) which Nausicaa's mules nibble on by the banks of the river (*Od.* 6.90). Cypress is sweet-scented, εὐώδης (*Od.* 6.564).

[41] The technological simile in the Polyphemos episode comparing the hiss of the olive stake as it is driven into the monster's eye to molten bronze dipped into water (*Od.* 9.391ff.) is a notable exception.

suggests why gates are said to "bellow." When Achilles spears Aineias' shield where the oxhide was lightest, λεπτοτάτη δ' ἐπέην ῥινὸς βοός (20.276), Aineias' shield cries shrilly upon impact: λάκε δ' ἀσπίς (20. 277).[42] Twelve lines earlier in Book 20 Aineias spears Achilles' shield and it bellows (μύκε, 260), like the bellowing cattle described on the Shield (18.580). When pierced, the animal by–product that has gone into making the shield is reanimated, as if to suggest that in the chaos of war the animal on the shield sympathetically feels the blow.

A sound sometimes is made a point of comparison in similes. The onomatapoeic word κλαγγή, or some verbal form thereof, is used nine times to describe the clamoring of birds, as in the short simile ἀμφὶ δέ μιν κλαγγὴ νεκύων ἦν οἰωνῶν ὥς (*Od.* 11.605).[43] The cry of birds in a simile easily harmonizes with a military description of shouting soldiers and clanking armor on the battlefield. In the famous Pygmy–crane simile (3.2ff.), the κλαγγή of the undisciplined Trojans setting out to battle is compared with the κλαγγή of cranes, in contrast to the stealth of the Achaians. The poet compares the grinding of bronze weaponry to the clatter made by the tusks of boars wheeling on huntsmen and dogs as they uproot the underbrush: πρυμνὴν ἐκτάμνοντες, ὑπαὶ δέ τε κόμπος ὀδόντων Ι γίγνεται ... ὥς τῶν κόμπει χαλκός ... (12.149–51, cf. 11.416f.).


### Interaction of Language Describing Motion and Emotion

Descriptions of movement are a key to understanding the poetic uses of animals in the poems. εἰλίπους, for example, captures the shambling gait of cattle, ταναύπους their long stride, ἀερσίπους the high–stepping of a horse. In the war epic, with its stress on speed and action, it is hardly surprising to find that animal similes serve to emphasize either mass or individual movement.[44] Animal motion imparts a sense of energy to the human counterpart. There are forty animal similes in both poems whose

---

[42] On one other occasion λάσκω describes the noise bronze makes when struck (14.25).

[43] In other places it describes the whining of pigs (*Od.* 14.412). Other onomatapoeic words like τρίζω describe the squeaking of bats (*Od.* 24.5), μυκάομαι the roar of a bull, (cf. μυκηθμός of oxen), μηκάομαι the bleating of sheep.

[44] Cf. M. Coffey, *AJP* 78 (1957) 121.

primary function is to highlight action. Many of the short animal similes re-volve structurally around two correlative verbs of action, e. g., σευάμε-νος ὥς θ' ἵππος ... ὥς 'Αχιλεὺς λαιψηρὰ πόδας ... ἐνώμα (22.22–24). The parallel verbs need not be contained within a simile but may straddle simile and narrative. A short simile comparing Meriones to a swooping vulture, αἰγυπιὸς ὥς (13.531), as he plucks the spear from the body of his victim, restates through the animal metaphor the warrior swooping (ἐπάλ-μενος) to spear him (15.529). In the internal development of similes verbs of motion are especially conspicuous. The psychological simile at 11. 548ff. comparing Aias to a frustrated lion comments primarily on the war-rior's disappointment after Zeus has halted his rampage. Nevertheless the simile bristles with movement. Farmers and dogs persist in their attempt to repulse the attack of a lion against a herd of cattle until the lion slinks away. There are five action verbs in this simile: ἐσσεύαντο, ἐλέσθαι, ἰθύει, πρήσσει, ἀίσσουσι, as well as the participle ἐσσύμενος. The oral tradi-tion is sensitive to the spirit, agility, and swiftness of animals, as the re-peated simile of a thoroughbred breaking free from his stall (6.506ff.=15. 263ff.) suggests. We share in the horse's taste of freedom as his pliant limbs effortlessly carry him forward to his favorite places.

Similes of swarming insects and migratory birds are especially well–suited to describing the mass–movement of armies.[45] In passages detailing the rout of armies κλονέω and φεύγω in the often complementary maraud-ing lion similes emphasize the fear and confusion of the victims. ἅλλομαι and ῥήγνυμι, among other verbs, depict the leap and mauling action of the lion. The phrase μακρὰ βιβάς/–άντα conveys the long stride of the lead ram (*Od.* 9.450) as well as the showy, athletic movements of warriors—for example, Aias springing from ship to ship (15.686, cf. 676), or Paris parading before Menelaos "like a lion" (μακρὰ βιβάντα ǀ ὥς τε λέων, 3.22f.). The verb ἐλίσσω, used of the boar swinging round to confront his attackers, allusively portrays the warrior parrying his pursuers. The ana-logical movement is pronounced in Aias' boar simile in the Patrokleia where he turns (ἐλέλιξεν, 17.278) on the Trojans and easily makes them scatter (ῥεῖα ... ἐκέδασσε, 17.285), as the boar wheeling (ἑλιξάμενος, 17.283) effortlessly disperses the hounds and huntsmen (ῥηϊδίως ἐκέ-δασσεν, 17.283). The parallelism between animal and human locomotion is so deeply embedded in the traditional language of the epic that even with-

---

[45] Scott, *Homeric Simile* 74f., 77–79; cf. Coffey, *op. cit.* 121, and Schnapp–Gourbeillon, *Lions* 28–32.

out the analogical device of the simile, animal movement is occasionally suggested by motion verbs in passages of heightened emotional intensity.[46] Several instances will be discussed below in more sustained analyses of hunting and herding imagery.

The emphasis on movement in animal similes is closely related to the complex of drives and emotions that bring to life animals, and by analogy, humans in the narrative.  Sarpedon's simile in the *Teichomachia* (12.299ff.), which evokes a lion "who for a long time has gone lacking meat, and the proud heart is urgent upon him I to get inside of a close steading and go for the sheepflocks ..." not only a describes the lion's movement but provides an explanation for it. The lion "evidently does so *because* he has a very pressing need of food and he sees there what will satisfy that hunger."[47] In an analogous manner Sarpedon's heart urges him to attack because he needs to cross the wall in order to win glory. In this and other lion similes, the lion's hunger is wedded to a psychological motivation expressed in the phrase "the proud heart is urgent upon him." Only once in a simile of a predator overcoming its prey is the physical motive alone given. In the simile at *Od.* 6.130ff. Odysseus' sexual appetite is expressed in terms of the lion's physical appetite (κέλεται δέ ἐ γαστήρ). Usually the instinctive drive of the animal is expressed as a heroic virtue, such as a proud or "very manly" (ἀγήνωρ) spirit. The motive for the animal in the similes is then used to explain the cause for human activity in the narrative.

Many of the developed animal similes, therefore, go beyond explaining events by physical needs and probe the psychological aspect in heroic terms. Animals in Homer are not presented in a purely naturalistic manner but symbolically as well. The emotive system of epithets and formulaic phrases revolves around distinguishing between animals as emblems of cowardice and bravery. In keeping with the concern in much of early Greek poetry to counterbalance praise and blame,[48] the majority of animal

---

[46] Cf. J. Pollard, *Birds in Greek Life and Myth* (1977) 157, on verbs that impart a bird–like motion to gods.

[47] M. C. Nussbaum, *The Fragility of Goodness* (1986) 267, where she goes on to explain that in terms of Aristotelian teleology human beings and animals alike move from point A to point B out of need for certain things once they see what will satisfy these needs. She observes that the Homeric animal simile relies on "a broad agreement in finding a common structure" for the reason for a given action. Nussbaum also discusses this simile in an interpretative essay in *Aristotle's* De Motu Animalium (1978) 59.

[48] Cf. G. Nagy, *The Best of the Achaeans* (1979) 222ff.

similes are either honorific or pejorative. An animal simile praises or casts shame on the object of comparison.

This usage can be traced to the ambivalence felt towards animals, as expressed in honorific epithets such as θυμολέων and in terms of abuse such as κυνῶπης. In the *Iliad* the difference between praise and blame in the animal metaphor is determined by the essential distinction between speedy motion and its opposite. Speed alone, however, does not automatically guarantee bravery, because a coward can flee quickly from the battlefield, just as cattle do in the similes when driven into a stampede, or deer when pursued by the hunter and his hounds. The deer not only flees but is sometimes paralyzed by fear. Lack of motion, however, can be a desirable quality when accompanied by staunch defense, the quality embodied by the boar.

According to the dramatic requirements of battle poetry the following equivalents based on motion and stillness are established in the animal imagery: the lion typically makes a swift, aggressive attack, while the boar stands his ground. Cattle flee out of fear, as do deer, fawns, and smaller birds. This dichotomy between cowardice and bravery is open to variations, as in the case of the lion, who may lose heart and retreat; but the basic distinction holds. This division generates the second system of epithets which emphasizes internal aspects.

These formulae are in turn closely allied with honorific epithets and terms of abuse. The boar is ὀλοόφρων, whereas deer are ἀνάλκιδες. Thus Achilles tells Agamemnon that he has the heart of a deer; a lion, on the other hand, has a θυμὸς ἀγήνωρ (24.42). Emotive terms often accumulate in the animal simile in order to amplify the raw forces of nature in the beast, and by analogy, in the human object. In Menelaos' emotional outburst against Euphorbos he appeals to Zeus to witness the warrior's hybris: "Father Zeus, it is not well for the proud man to glory. I Neither the fury of the leopard is such, not such is the lion's, I nor the fury of the devastating wild boar, within whose breast I the spirit is biggest and vaunts in the pride of his strength, is so great I as goes the pride in these sons of Panthoös of the strong ash spear" (17.19–23).

An example from Achilles' *aristeia* indicates how emotive terms and action verbs in a simile interact with the surrounding narrative to mirror the vicissitudes of Agenor, the hero's last victim before encountering Hektor. When Apollo rouses Agenor to encounter Achilles, the god stands beside him leaning on an oak. Agenor, aware of a divine presence, grows angry and speaks to his own spirit (21.553–62). At the conclusion of Agenor's speech ensues a leopard simile (21.573–78).

> ὤ μοι ἐγών· εἰ μέν κεν ὑπὸ κρατεροῦ ᾽Αχιλῆος          21.553
> φεύγω, τῇ περ οἱ ἄλλοι ἀτυζόμενοι κλονέονται,
> αἱρήσει με καὶ ὥς, καὶ ἀνάλκιδα δειροτομήσει.          555
> εἰ δ᾽ ἂν ἐγὼ τούτους μὲν ὑποκλονέεσθαι ἐάσω
> Πηλείδῃ ᾽Αχιλῆι, ποσὶν δ᾽ ἀπὸ τείχεος ἄλλῃ
> φεύγω πρὸς πεδίον ᾽Ιλήιον, ὄφρ᾽ ἂν ἵκωμαι
> ῎Ιδης τε κνημοὺς κατά τε ῥωπήια δύω·
> ἑσπέριος δ᾽ ἂν ἔπειτα λοεσσάμενος ποταμοῖο          560
> ἱδρῶ ἀποψυχθεὶς προτὶ ῎Ιλιον ἀπονεοίμην·
> ἀλλὰ τίη μοι ταῦτα φίλος διελέξατο θυμός;

cf.

> ἠύτε πάρδαλις εἶσι βαθείης ἐκ ξυλόχοιο          573
> ἀνδρὸς θηρητῆρος ἐναντίον, οὐδέ τι θυμῷ
> ταρβεῖ οὐδὲ φοβεῖται, ἐπεί κεν ὑλαγμὸν ἀκούσῃ·          575
> εἴ περ γὰρ φθάμενός μιν ἠὲ βάλῃσιν,
> ἀλλά τε καὶ περὶ δουρὶ πεπαρμένη οὐκ ἀπολήγει
> ἀλκῆς, πρίν γ᾽ ἠὲ ξυμβλήμεναι ἠὲ δαμῆναι·
> ὡς ᾽Αντήνορος υἱὸς ἀγαυοῦ, δῖος ᾽Αγήνωρ,
> οὐκ ἔθελεν φεύγειν, πρὶν πειρήσαιτ᾽ ᾽Αχλιῆος          580

The simile represents the leopard emerging from the wooded thicket to face the hunter who has been tracking her. The leopard does not fear or even contemplate flight "even though one be too quick for her with spear thrust or spear thrown | struck with the shaft though she be she will not give up her fighting | fury, till she has closed with one of them or is over-thrown; | so the son of proud Antenor, brilliant Agenor, | was unwilling to run away until he had tested Achilles ..." The simile both looks backward to Agenor's speech as it specifically contrasts his present action with his previous hesitation as well as announces what the audience knows is about to happen: one more skilled with the spear will soon destroy him. In action he is comparable to the leopard; in a state of hesitation, he is like a fright-ened animal taking cover in a thicket.

Although there is no explicit animal imagery in Agenor's speech, we may observe in these lines certain linguistic details from animal similes. The verbs describing stampede and flight (κλονέονται 554, ὑποκλονέ-

εσθαι 556, φεύγω 554, 558) all occur elsewhere in marauding lion similes. κλονέω properly portrays the lion setting the cattle into stampede
(15.324). Agenor's speech evokes a mountain setting: ῥωπήια (559), the
word for the undergrowth where Agenor would hide, is found in a simile
of lions seizing carcasses from dogs, then carrying them into the underbrush (13.198ff.). The mountainous locale, and specifically a wooded lair
(ξύλοχος, 573), persists in the ensuing leopard simile honoring Agenor's
decision not to flee but to confront Achilles. The simile is anticipated by the
action verb in its participial form (ἀλείς, 571) which means "crouching to
spring." The same verb depicts the lion's movement in Achilles' lion simile
in his *aristeia* (20.168). The animal image evoked in Agenor's speech belongs to an accumulation of images that compare the Trojans in Books 21
and 22 to frightened fawns, creatures which are given to flight or paralysis,
since they have no spirit (ἀλκή, cf. 4.243–45).[49] At 21.29 the twelve
Trojans chosen by Achilles as restitution for Patroklos are led out of the
water "bewildered like fawns." The participle τεθηπότες occurs elsewhere
of fawns in Agamemnon's rebuke to his men (4.243ff.);[50] a similar verbal
form, πεφυζότες, at the beginning of Book 22 modifies the Trojans, who
are likened to fawns as the bewildered men slake their thirst. Yet another
similar phrase, ἀτυζόμενοι (21.554), serves as the supplementary participle of κλονέω in Agenor's speech. πεφυζότες modifies the Trojans
twice (21.528, 532) prior to Agenor's speech, both times in interaction
with the verb κλονέω. Priam looks out from the bastion and sees Achilles,
" ... where before him I the Trojans fled in the speed of their confusion, no
war strength left them." He urgently advises, "Hold the gates wide open in
your hands, so that our people I in their flight can get inside the city, for
here is Achilles I close by, stampeding them, and I think there will be disas-
ter."

The accumulation of vocabulary for action and inaction can be related
to the dilemma Agenor expresses in his speech and which he resolves by
activity. Agenor chooses to distinguish himself from the other Trojans,
who behave in a cowardly manner, like fawns or cattle. He will not be, as
he momentarily pictures himself, a terror–stricken creature cowering in the
underbrush, but rather the leopard who mounts an attack from the thicket.
This example of sustained animal imagery is part of a larger tapestry of
hunting and herding imagery discussed in Chapters 4 through 6.

---

[49] Cf. pp. 89–90.

[50] cf. Kirk, *Commentary, ad loc.*

# Chapter III
# The Beast of War: Lion Similes

## Typology[1]

We have noted the heightened descriptive quality of language for animals in the background of the poems and have observed the interaction of animal imagery in simile and narrative. We now turn to examine how the principal simile type behaves from the point of view of its internal scene and how it functions in relation to its context.[2] We will see that all lion similes are basically modeled on a single type (the marauding lion) involving the confrontation of men and dogs with a wild beast, and that this con-

---

[1] The formulaic language in lion similes is included in Appendix C.

[2] There are 50 lion similes altogether, including θήρ similes, and more hexameter verses (146) occurring in lion similes than in any other simile type. θήρ in the singular, and once in the dual, designates the lion in similes. This is readily apparent at 15.633, where θήρ in a simile is simply a synonym for λέων (15.630). Therefore θήρ similes, of which there are five, will be included with other lion similes. For a list of line references to lion similes see Appendix D. Lion similes in the *Odyssey* are discussed by Moulton, *Similes* 139–41; cf. R. Friedrich, *AJP* 102 (1981) 120–137 and W. T. Magrath, *CJ* 87 (1982) 205–212. Outside of similes the lion appears in proper names, as an epithet, in material items, in ecphrastic descriptions of works of art, as a metaphor for Artemis, and in metamorphoses in the Circe and Proteus episodes. The stem λεοντ- from the accusative of λέων appears in the name of the warrior Leonteus, cf. Δημολέων 20.395. λέων occurs in the compound θυμολέων. Four warriors merit the epithet, Tydeus, Herakles, Achilles, and Odysseus in the *Odyssey*. In the case of Odysseus the epithet corresponds to the somewhat emblematic associations between the hero and the lion as revealed in the similes in the *Nausicaä* and the *Mnesterophonia*. In *Od.* 4 Penelope refers to her husband as θυμολέων before and after her dream; her sense of frustration is compared with that of a hemmed–in lion (*Od.* 4.791, cf. *Od.* 4.724, 814.). Diomedes, who along with Odysseus is compared to a pair of lions in a simile in the *Doloneia*, sets out on the night–escapade wearing a lion–skin (10.297f.; 177). Nestor also wears a skin (10.23). Lions appear on Herakles' baldric in the Underworld scene (*Od.* 11.611) and on Achilles' shield (18.579ff.). Metaphorically Apollo says that Zeus sent Artemis to be a "lioness among women" (21.483f.), since she is traditionally the goddess responsible for the death of women. Achilles, for example, wishes that Artemis had dealt Briseis one of her arrows (19.59f., cf. 6.205). In a mythical setting some of Circe's former victims have undergone leonine metamorphosis (*Od.* 10.212, cf. 218), and Proteus so transforms himself (*Od.* 4.356). The lion takes pride of place in the chimaira hybrid (6.181).

figuration, which reproduces a warrior's encounter on the battlefield, is the common source for the boar hunting similes, as demonstrated in Chapter 5.

The simile at 20.164ff., the longest in the *Iliad*,[3] may serve as our model.

> Πηλεΐδης δ' ἑτέρωθεν ἐναντίον ὦρτο λέων ὥς          20.164
> σίντης, ὅν τε καὶ ἄνδρες ἀποκτάμεναι μεμάασιν
> ἀγρόμενοι πᾶς δῆμος· ὃ δὲ πρῶτον μὲν ἀτίζων
> ἔρχεται, ἀλλ' ὅτε κέν τις ἀρηϊθόων αἰζηῶν
> δουρὶ βάληι ἐάλη τε χανών, περί τ' ἀφρὸς ὀδόντας
> γίγνεται, ἐν δέ τέ οἱ κραδίηι στένει ἄλκιμον ἦτορ,
> οὐρῆι δὲ πλευράς τε καὶ ἰσχία ἀμφοτέρωθεν          170
> μαστίεται, ἑὲ δ' αὐτὸν ἐποτρύνει μαχέσασθαι,
> γλαυκιόων δ' ἰθὺς φέρεται μένει, ἤν τινα πέφνηι
> ἀνδρῶν, ἢ αὐτὸς φθίεται πρώτωι ἐν ὁμίλωι
> ὣς 'Αχιλῆ' ὄτρυνε μένος καὶ θυμὸς ἀγήνωρ
> ἀντίον ἐλθέμεναι μεγαλήτορος Αἰνείαο.          175

The simile occurs in Achilles' *aristeia* as Aineias presents himself as the first opponent. The actual confrontation is postponed until the end of the Book, but the simile has the important function of mobilizing Achilles as an active fighter (cf. Krischer's *Auszug zum Kampf*), as well as characterizing the bestial fury that motivates his revenge. Throughout the simile we may observe an alternation of verses or phrases relating to motion and physical states on the one hand, and to emotive states on the other.

The lion is described as a ravening beast, σίντης, one who has been a threat to the community. All the inhabitants, πᾶς δῆμος, gather to repulse the beast. At first the lion advances heedless of danger, but then one of the assembled party strikes with a spear. The lion's anger rises to a peak; he is on the verge of attack. A full five lines describe the physical manifestations of anger: his mouth gapes as foam oozes through the fangs; he whips himself into action, lashing his tail against the ribs and flanks, and the eyes glower as he hurls himself forward for a head–on attack: γλαυκιόων δ' ἰθὺς φέρεται μένει (20.172). The concluding phrases of the simile, ἤν τινα πέφνῃ | ἀνδρῶν ἢ αὐτὸς φθίεται πρώτῳ ἐν ὁμίλῳ (172f.), restate in the animal metaphor a heroic ideal: in the face of attack one must risk fighting no matter what the consequences. The simile is composed in ring–

---

[3] I.e., 10 lines; the nightingale simile of 13 lines at *Od.* 20.66ff. is the longest Homeric simile.

form. It begins with an introductory action verb in the narrative (ὦρτο) to describe Achilles' physical movement and concludes with a synonym (ὄτρυνε) telling of the arousal of Achilles' inner forces, his μένος and θυμός. The symmetry is sharpened by the repetition of (ἐν)αντίον (164, 175). In the introduction the lion is announced, and in the conclusion Achilles is named. The description of the lion's intent to attack and feast on cattle alternates with an explanation of the defenders' intent to counterattack. We next hear of the fearless character of the lion. The simile then shifts to an account of the actual attack by the defenders and back again to a characterization of the lion's furor and physical manifestation of anger (foaming mouth, tail lashing rib–cage, etc.), and back once again to the counterattack. At this point the poet's remark that the lion risks attack in spite of all odds injects a note of pathos to balance with the earlier description of the lion's ethos. The ring is then completed by the resumptive clause summarizing the warrior's corresponding intent to attack. Although attack and counterattack are mentioned in the simile, the outcome of the confrontation is left in suspense in the alternative clauses ἤν τινα πέφνῃ | ἀνδρῶν, ἢ αὐτὸς φθίεται... In the same way we are left in suspense about the confrontation of Achilles and Aineias by the long verbal exchanges that follow.

Although the simile is typical, it is, like other similes, adapted to the extraordinary circumstances of the moment when Achilles finally breaks his self–imposed withdrawal from the fighting; its very length and wealth of internal detail reflect its import at this critical juncture in the plot. ἀτίζων, for example, a Homeric *hapax* which portrays the lion as aloof from his opponents, nicely corresponds with the sovereign character of the hero. The elaborate description of the lion's anger is inspired by the wrath theme. The simile resembles the portion of the Meleager–paradigm reporting the circumstances leading up to Meleager's efforts to organize a party to exterminate the boar (9.538–49). Like the boar and indeed like Achilles in his extra–Trojan deeds (20.90ff., 188ff.), the lion to whom Achilles is compared has for some time disrupted the community; like Meleager and the farmers in the simile, Aineias now gathers the Trojans to repel him.[4]

Our model simile describes the marauding lion. Out of a total of 50 lion similes, twenty–seven portray a lion or lions attacking domesticated animals, all except one occurring in the *Iliad*. Nineteen of these mention the

---

[4] This simile is discussed in relation to the plot, pp. 88–89.

efforts of the herdsmen and their dogs to repel the invader. Marauding lions are by no means incontestably victorious. Two of them die; twice the beast's death is given as a possible outcome; and three times the lion turns cowardly and slinks away.[5] In addition to the marauding type, seven developed similes represent the lion attacking various wild animals, including the wild goat, deer, and the wild boar.[6] There are nineteen similes with no remarkable scene or action, e.g., βαῖνε, λέων ὥς, ἀλκὶ πεποιθώς (5. 299).

A glance at the schematic analyses of lion and boar similes in this and the following chapter will show that basic components of the model simile are repeated in less elaborate forms in the developed lion similes. Sometimes the sequence of events is altered, and sometimes the motive of the human attackers is suppressed, but the basic ring form and the alternating points of view in the internal narrative persist.

### Anatomy

The internal scene of marauding lion similes outlines the physical setting, the lion's habitat, the farmers and dogs, and the predator. The background of lion similes is variable and sketchy, as in most Homeric topographical or geographical descriptions.[7] αὐλή (5.138[8]), μέσσαυλος (17.657, etc.), σταθμός (5.557, etc.) and πυκινὸς δόμος (12.301, Od. 6.134), loosely refer to the fold or area in which the sheep or cattle are kept. One could hardly expect a more elaborate picture of what must have been a humble enclosure. The verbs δεύω, (μεθ)άλλομαι, πρήσσω and οἰμάω imply that the lion penetrates a barricade. Alternatively the lion at-

---

[5] Lion dies: 5.558, 12.46; death mentioned: 12.305f., 16.748, 20.172f.; cowardly lion: 11.555, 17.11f., 664.

[6] Whereas the standard scene in Homer portrays a lion marauding domesticated animals (cow and sheep), in Greek art from Geometric times onward the recurring scene depicts a lion mauling a wild animal (deer or fawn); cf. H.–G. Buchholz et al., *Jagd und Fischfang, Archaeologia Homerica* J (1975) 27–30, and G. E. Markoe, "The 'Lion Attack' in Archaic Greek Art: Heroic Triumph," *CA* 8 (1989) 868–92.

[7] Cf. P. Vivante, *The Homeric Imagination* (1975) Ch. III, "On the Representation of Nature and Reality in Homer," esp. 72–79.

[8] Shipp, *Studies* 245, interprets αὐλή here as a wall, cf. the Cyclops' cave, *Od.* 9.184 and Eumaios' steading, *Od.* 14.5.

tacks herds of grazing cattle.[9] The lion inhabits the mountains (e.g., 5.554f.); the distinction between the semi–civilized territory of the shepherd and the savage haunts of the lion emerges most clearly in a simile in which hill–bred lions plunder the σταθμοὺς ἀνθρώπων (5.557), where the emphasis lies on ἀνθρώπων, cf. μῆλα βροτῶν (24.43).[10] Only the barest details concerning the lion in his natural habitat emerge. The lion lives in a ξύλοχος, a term also meaning "thicket."[11] The details of the physical setting are not precisely correlated with tactical situations in the narrative, except that the cattle steading loosely becomes a metaphor for the Trojan citadel or the Achaian wall.

An impressive variety of phrases delineate the human actors in the similes: βώτορας ἄνδρας (12.302), ποιμήν (5.137), ἀνδρῶν ἠδὲ κυνῶν (10.186), κύνες τε καὶ ἀνέρες (11.549=15.272), κύνες τε καὶ ἄνδρες (17.110), κύνες τ' ἄνδρες τε νομῆες (17.65), ἄνδρες ἐπακτῆρες (17.135), ποιμένες ἄγραυλοι (18.162) and αἰζηοί (3.26, 18.581, 20.167). This noteworthy variety of phraseology for what is a recurrent semantic idea can be observed in other features of the lion similes. Sometimes the men are specifically said to be absent: ἀσημάντοισιν or σημάντορος οὐ παρεόντος (10.485, 15.325). Once the shepherd is called "inexperienced," νομεὺς οὔ πω σάφα εἰδὼς | θηρὶ μαχέσσασθαι (15.632f.). The poet singles out the action of the defenders throwing spears, θοῆς ἀπὸ χειρὸς ἄκοντι (12.306) or θαμέες γὰρ ἄκοντες | ἀντίον ἀίσσουσι θρασειάων ἀπὸ χειρῶν (11.552f.), because this is what warriors do. Variations on this basic activity occur. καιόμεναί τε δεταί (11.554) refers to attempts to intimidate the lions with torches. In a simile from Menelaos' *aristeia* that conveys a clear feeling for the fear and helplessness of the defenders, the farmers try unsuccessfully to frighten off the beast with their cries, and no doubt, the baying of dogs (17.66f.):

πολλὰ μάλ' ἰύζουσιν ἀπόπροθεν οὐδ' ἐθέλουσιν
ἀντίον ἐλθέμεναι μάλα γὰρ χλωρὸν δέος αἱρεῖ.

---

[9] E.g., ξύλοχον κατὰ βοσκομενεάων, (5.162, cf. 17.62ff.) goats, (11.383, 16.487ff.); on the Shield the cattle are grazing in a meadow by a resounding river: 18.576ff.

[10] Cf. C. Macleod, *Homer*, Iliad *XXIV* (1982) *ad loc.*, "A human characteristic is here transferred to the animal ... For the same reason βροτῶν is no filler."

[11] 5.162, *Od.* 4.335; cf. 21.573, of the haunts of the leopard, where the term does not specifically designate "lair" as in the *Odyssey* passage.

In the model simile we saw the determination of the community to repulse
the noxious beast through a concerted group effort. But the lion may catch
a solitary defender off guard. Once a frightened herdsman gives up fighting
and cowers in the steading, ἀλλὰ κατὰ σταθμοὺς δύεται (5.140). The
convincing insights into the reactions of the defenders argue in favor of a
first–hand acquaintance with the dangers which the invader poses to the
herdsman. In contrast to the aggressive hounds of the boar similes, the
watchdogs are notably inactive and do not receive distinctive epithets.

The lion is very much a three–dimensional figure in the similes. A
more convincing composite portrait emerges for the lion than for any other
animal in the similes. Of the body of the lion, only the sheerest outlines
emerge in the model simile mentioning the ribs and flank, πλευράς τε καὶ
ἰσχία (20.170). In that simile the details of the lion's anatomy remain
hazy: the tail lashes the rib–cage, the lion bares its foam–drenched fangs,
and the eyes gleam. Variations on the last two details account for all other
physical descriptions. Five similes refer to the deadly grip of the lion's
jaws, κρατεροῖσιν ὀδοῦσιν (11.114, 175=17.63), or γαμφηλῇσι(ν)
(13.200, 16.489). The belief that the lion's eyes blaze recurs at *Od.* 6.
131f.: ἐν δέ οἱ ὄσσε | δαίεται (cf. χαροποί, *Od.* 11.611). The phrase
ἐπισκύνιον κάτω ἕλκεται ὄσσε καλύπτων (17.136) contains a striking
bit of detail. ἐπισκύνιον refers to the furl of skin above the lion's eyes. In
the simile the lioness leads her young through the forest when huntsmen
appear by chance. She stands her ground and holds them with their gaze.
The expression may be a leonine version of ὑπόδρα ἰδών "looking darkly
under the brow," and occurring at 17.169.[12]

The epithet ἠυγένειος (with λίς 15.275, 17.109, 18.318; with λέων
*Od.* 4.456) probably means "well–bearded," i.e., having an impressive
mane. Common sense would then suggest a reference to the lion as opposed to the lioness, though such an inference runs into problems in the
simile at 18.318ff. of a lion(ess?), λὶς ἠυγένειος, bereft of young. αἴθων
is the only color epithet. The young of the lioness are referred to as τέκνα
(11.113, 17.133), which serves equally to designate the young of wasps,
sparrows, etc. The cubs are also called σκύμνοι. As Stanford states in his
commentary to the *Odyssey*, "Homer's lions never roar; indeed they are
silent in Greek literature till the *Hymn to Aphrodite* (159) calls them

---

[12] Cf. J. P. Holoka, "'Looking Darkly' (ΥΠΟΔΡΑ ΙΔΩΝ)," *TAPA* 113 (1983); for
the mesmerizing gaze of warriors and animals see S. H. Lonsdale, *CJ* 84 (1989) 325–
333.

βαρύφθογγοι."[13] One could argue that the oral tradition was ignorant about the habits of lions and thereby missed an excellent opportunity to draw a parallel between the battle–cry of the warrior and the roar of the lion. One could also argue that silence is in keeping with the portrayal of the lion as a stealthy creature. In a simile we hear that the beast stalks his prey at night μολὼν ἐν νυκτὸς ἀμολγῷ (11.173), and in another at 10. 183ff. the watchdogs hear the beast, θηρὸς ἀκούσαντες κρατερόφρονος, stealing through the forest. The only auditory description in a lion simile relates to the groan of an internal organ. The anthropomorphizing phrase, ἐν δέ τέ οἱ κραδίη στένει ἄλκιμον ἦτορ (20.169), alludes to Achilles' anger, much as Odysseus' "barking" heart resonates with the simile of a bitch protecting her puppies (*Od.* 20.13ff.).

Lion similes in general abound in expressions conveying emotional or mental states. These words and phrases reflect or sharpen the emotion of the subject of comparison, at the same time as they animate the beast himself. Some expressions have the nature of an epithet, while others evince a distinct emotion which elucidates a point in the narrative. ἀλκὶ πεποιθώς in end–line position at 5.299, 17.61, or σθένεϊ βλεμεαίνων (–ει), an epithet of Hektor and also of the boar (12.42 [lion–boar], 17.135), show how straightforwardly the beast and the mighty warrior share heroic status. Several expressions refer to the seats of emotion or intellect, with characteristic psychological inexactitude. κρατερόφρονος (10.184), μεγάθυμος (16.488), and μέγα φρονέοντε (16.758, 824) characterize the lion's confidence in attack. κακὰ φρονέων (10.456) implies evil intent (cf. ὀλοόφρων 15.630, also used of the boar). ὡς ἄγρια οἶδεν (24.41) denotes the lion's disposition for savagery.

Distinct emotions take hold of the lion. He feels joy in coming upon a carcass (ἐχάρη, 3.23); the lion contending with the boar over a mountain spring vanquishes his opponent χάρμῃ. A bitter anger seizes the lioness bereft of her young (δριμὺς χόλος αἱρεῖ, 18.322). These are examples of emotions which correspond directly to those of the subjects of comparison. Words such as χαίρω, χόλος, and χάρμη have strong heroic overtones that acquire a refreshing ring when recast in the animal metaphor.

---

[13] Stanford, Commentary on *Od.* 4.335ff; T. Jahn, *Zum Wortfeld "Seele–Geist" in der Sprache Homers, Zetemata* 83 (1987) 234, observes that the lion does not groan audibly but that his ἦτορ does so inwardly, in his κραδίη.

Lions are fearful and cowardly as well as courageous. The lion's valorous heart freezes within his breast as he retreats (17.109ff.), like the lion who slinks away at dawn disappointed in his heart (11.555=17.664, cf. 16.753 ὤλεσεν ἀλκή). A crowd of men drives a circle around a fearful lion who contemplates his next move (μερμήριξε λέων ... δείσας, *Od.* 4.791f.). The poet's sympathy with the weaknesses of the lion complements his admiration of his strengths, and the net result is one of heightened realism.


## Action in Lion Similes

The convincing nature of the lion simile emerges most strongly in descriptions of the lion's movements. A score of different verbs are used to choreograph the lion's preparatory maneuvers—the entry into the steading, the stampeding of herds, or the pernicious crouch before the leap. Another dozen verbs tell of his attack and the subsequent feast. The wide assortment of verbs rounds out the contours of the beast. Their descriptive flavor infuses him with life and animates him with vicious, athletic movements.

Compounds of ἄλλομαι specify the leap over the barricade. Note the precision in the simile at 5.136ff. When he approaches he jumps over: αὐλῆς ὑπεράλμενον (5.138); once he has made his kill he leaps out of the "deep yard" ἐξάλλεται (5.142, cf. ἄλτο 16.755, μετάλμενος 12.305). κεραΐζων (16.752), which depicts the lion ravaging his victims, also denotes the act of plundering a city (16.830, 24.245). Twice the aorist middle of ὄρνυμι immediately precedes the beginning of a lion simile, as if to launch the warrior into action in his *aristeia* and to signal that the beast is attacking (11.129, 20.164). ὀρούω (15.635), which describes the onslaught of the predator, is also used for the flight of the weapon (13.505=16.615) and the attack of the warrior. Portrayals of physical action alternate with explanations of internal motivation. The lion is said to make his entry for various reasons: his spirit or appetite bids him (κέλεται, 12.300, *Od.* 6.133), he longs for his favorite meat κρειῶν ἐρατίζων (11.551=17.660), he is eager or wants "to make a try," ἐμμεμαώς (5.142), ἐσσύμενος (11.554=17.663), πειρητίζων, πειρήσοντα (12.47, 301). The last two are echoes of the warrior's appetite for fame (e.g., 15.615, 21.580).

The noun οἶμα denotes the calculated leap. The parallelism between the leap of the lion and that of the warrior is made explicit by the ring that frames a simile addressed by the poet to Patroklos: οἶμα λέοντος ἔχων ... ὥς ... Πατρόκλεες ἆλσο μεμαώς (16.752ff). Then in the next line Hektor becomes the subject of the verb ἆλτο, as he too becomes like a lion in another simile beginning in the next line. Eventually he will overtake Patroklos, and his lion image, as a lion defeats a boar (16.823ff.). The verb οἰμάω, used to describe the swoop of the eagle, does not occur with lions in Homer. Instead the poet uses θρῴσκω (5.161), and the expression ἰθὺς φέρεται (20.172, cf. ἰθύσῃ 12.48). Numerous expressions portray the actual attack. (συν)έαξε (11.114, 175), ἁρπάζω (17.62, 12.305) ἔπεφνε (16.487), ἔδει, δάπτει, and the formulaic phrase λαβὼν κρατεροῖσιν ὀδοῦσι(ν) (11.114, 175, 15.636, 17.63) portray him sinking his teeth into the carcass. Several times the lion is turned away from the steading by the men. The expression ἐντροπαλιζόμενος, ὀλίγον γόνυ γουνὸς ἀμείβων (11.547, cf. 17.109) is an accurate representation of how a lion will slowly back off, always keeping his gaze turned towards his opponents until he dares to turn and make his getaway. κλονέω depicts the lion setting the cattle into stampede (15.324). The active of φοβέω likewise denotes the beast throwing the herds into panic (11.173); in the same simile one may note the passive voice for the flight of the cow and the human objects of comparison, the Trojans. The use of φοβέω in lion similes or in the resumptive clause (5.140, 11.121, 11.172, 15.326, 637) reflects the important structural function which certain lion similes have in marking the rout of warriors, one of the two principal battle situations most frequently juxtaposed with similes, as discussed in the following chapter.[14]

---

[14] The details in marauding lion similes find close parallels in modern accounts of cattle herding among the Masai tribes of East Africa, where lions are a very real threat; cf. Guggisberg, *Simba, the Life of the Lion* (1961) 122. For the accuracy of Homer's description of the lion's method of attack, see H. W. Auden *CR* 10 (1896) 107.

# Chapter IV
# The Lion as Marauder of the Herds
# and as Hunter

Having examined the pictorial and linguistic components of lion similes, we may make some preliminary observations about how similes behave when deployed by the poet in the narrative. We will begin with similes which exhibit relatively simple and straightforward connections with
their narrative environment and then progress to more elaborate instances
of lion imagery sustained over a wider stretch of the narrative. At the outset
two brief remarks may be made about the structural and emotive function
of lion similes. There are basically two types of confrontations in battle
scenes: man to man combat (cf. Krischer's *Einzelkämpfe* or *Monomachie*)
and the mêlée, often followed by a rout where a major warrior turns the
massed enemy to flight (cf. Krischer's *Ansturm auf die Phalangen*, *Flucht*,
and *Verfolgung des Heeres*). The simile of two lions fighting over a stag
and the lion vs. boar simile at the end of the *Patrokleia* are well–known examples of lion similes which underscore man–to–man combat. The second
developed lion simile from the *aristeia* of Agamemnon, to be discussed
below, emphasizes rout. The lion similes in Books 17 and 18 will not be
discussed as a group in the present chapter but will be treated in the analysis of hunting imagery in Chapter 6.

A lion simile from the *Odyssey* describes the Cyclops' eating habits
(*Od.* 9.292f.):

ἤσθιε δ' ὥς τε λέων ὀρεσίτροφος, οὐδ' ἀπέλειπεν,
ἔγκατά τε σάρκας τε καὶ ὀστέα μυελόεντα

In this case, there is no linguistic correspondence between simile and narrative. The simile functions almost like an adverb to emphasize the uncivi-
lized manner in which the monster devours his human victim. The short
simile from the ransom scene, Πηλείδης δ' οἴκοιο λέων ὥς ἆλτο θύ-
ραζε (24.572), likewise has little or no connection with the narrative, except that the narrative verb ἆλτο describes the manner in which Achilles

sprang into Priam's view and caused the old man alarm. The short similes
at 5.299 and 7.256 provide further examples of lion similes showing min-
imal contact with their surroundings.

Τρωσὶν μὲν προμάχιζεν 'Αλέξανδρος θεοειδὴς          3.16
παρδαλέην ὤμοισιν ἔχων καὶ καμπύλα τόξα
καὶ ξίφος· αὐτὰρ δοῦρε δύω κεκορυθμένα χαλκῶι
πάλλων 'Αργείων προκαλίζετο πάντας ἀρίστους
ἀντίβιον μαχέσασθαι ἐν αἰνῆι δηιοτῆτι.              20
Τὸν δ' ὡς οὖν ἐνόησεν ἀρηίφιλος Μενέλαος
ἐρχόμενον προπάροιθεν ὁμίλου μακρὰ βιβάντα,
ὥς τε λέων ἐχάρη μεγάλωι ἐπὶ σώματι κύρσας
εὑρὼν ἢ ἔλαφον κεραὸν ἢ ἄγριον αἶγα
πεινάων· μάλα γάρ τε κατεσθίει, εἴ περ ἂν αὐτὸν     25
σεύωνται ταχέες τε κύνες θαλεροί τ' αἰζηοί·
ὡς ἐχάρη Μενέλαος 'Αλέξανδρον θεοειδέα
ὀφθαλμοῖσιν ἰδών· φάτο γὰρ τίσεσθαι ἀλείτην·
αὐτίκα δ' ἐξ ὀχέων σὺν τεύχεσιν ἆλτο χαμᾶζε.

Here a precise homologation exists between ἐχάρη in 23 and ἐχάρη
in 27.[1] The repeated word, as often, describes an inner state shared between man and animal, such that we simultaneously experience the motives
of the respective attackers. The internal scene of the simile matches the narrative situation only insofar as the sight of Paris and the animal carcass
satisfy the appetites of Menelaos and the lion. The feline subject of the
simile is triggered by the leopard skin of Paris (17), here a sign of luxury,
not bravery. The inversion that then makes Menelaos comparable to the
lion and Paris to the victim is typical of ironic reversals between simile and
narrative. Linguistically there is more than meets the eye. The verb ἆλτο
of the resumptive clause carries over the image of the lion by suggesting a
further comparison between Menelaos' leap from the chariot (329) and the
leap of a lion.


## The *Aristeia* of Diomedes: Book 5

There are six lion similes, three short and three developed, in Book 5.
In addition Tlepolemos invokes the name of his father, Herakles, who is
accorded the epithet θυμολέων (639).  As we will see in the case of the *ari-*

---

[1] Cf. Moulton, *Similes* 89.

*steia* of Agamemnon, animal imagery in Book 5 is restricted to the lion.[2] After Diomedes suffers a wound to the shoulder by Pandaros' arrow he prays to Athene to stand by his side now "if ever before in kindliness you stood by my father" so that he may avenge himself (116f.). Athene reassures him, "Be of good courage now, Diomedes, since I have put inside your chest the strength (μένος) of your father untremulous, such as the horseman Tydeus of the great shield had" (124–26).

Refortified, Diomedes enters the foremost ranks, more eager than before, "and a force (μένος) three times as strong took hold of him (136)." The first lion simile, beginning in the fifth foot of this verse, marks the warrior's outward physical reanimation which is accomplished by an internal transformation. Athene conjures up and imparts to Diomedes Tydeus' μένος, which metaphorically takes the form of the lion simile. The second lion simile (161ff.) brackets and recapitulates a description of Diomedes killing three pairs of warriors.

> ἣ μὲν ἄρ' ὣς εἰποῦσ' ἀπέβη γλαυκῶπις Ἀθήνη,          5.133
> Τυδείδης δ' ἐξαῦτις ἰὼν προμάχοισιν ἐμίχθη
> καὶ πρίν περ θυμῶι μεμαὼς Τρώεσσι μάχεσθαι·
> δὴ τότε μιν τρὶς τόσσον ἕλεν μένος ὥς τε λέοντα,
> ὅν ῥά τε ποιμὴν ἀγρῶι ἐπ' εἰροπόκοις ὀίεσσι
> χραύσηι μέν τ' αὐλῆς ὑπεράλμενον οὐδὲ δαμάσσηι·
> τοῦ μέν τε σθένος ὦρσεν, ἔπειτα δέ τ' οὐ προσαμύνει,
> ἀλλὰ κατὰ σταθμοὺς δύεται, τὰ δ' ἐρῆμα φοβεῖται·          140
> αἳ μέν τ' ἀγχιστῖναι ἐπ' ἀλλήλῃσι κέχυνται,
> αὐτὰρ ὃ ἐμμεμαὼς βαθέης ἐξάλλεται αὐλῆς·
> ὣς μεμαὼς Τρώεσσι μίγη κρατερὸς Διομήδης.
> Ἔνθ' ἕλεν Ἀστύνοον καὶ Ὑπείρονα ποιμένα λαῶν,
> τὸν μὲν ὑπὲρ μαζοῖο βαλὼν χαλκήρεϊ δουρί,          145
> τὸν δ' ἕτερον ξίφεϊ μεγάλωι κληῖδα παρ' ὦμον
> πλῆξ', ἀπὸ δ' αὐχένος ὦμον ἐέργαθεν ἠδ' ἀπὸ νώτου.
> τοὺς μὲν ἔασ', ὃ δ' Ἄβαντα μετώιχετο καὶ Πολύειδον

---

[2] One short and two developed lion similes (476: 136ff., 161f.) refer to Diomedes, the main character of the Book (476 allusively). The two developed lion similes can be taken together in sequence since the second resolves the incomplete action of the first. One developed and one short simile (554ff., 782) refer to Diomedes' men; Aineias is the object of comparison in the remaining short simile (299). Moulton, *Similes* 58–64, discusses the similes in Book 5; cf. Schnapp–Gourbeillon, *Lions* 100–104; A. Thornton, "The *Aristeia* of Diomedes as the Main Theme in the Composition of Books 5–8 of the *Iliad*," *AULLA* 18, Congress Proceedings (1977).

υἱέας Εὐρυδάμαντος ὀνειροπόλοιο γέροντος·
τοῖς οὐκ ἐρχομένοις ὃ γέρων ἐκρίνατ' ὀνείρους,    150
ἀλλά σφεας κρατερὸς Διομήδης ἐξενάριξε·
βῆ δὲ μετὰ Ξάνθόν τε Θόωνά τε Φαίνοπος υἷε
ἄμφω τηλυγέτω· ὃ δὲ τείρετο γήραϊ λυγρῶι,
υἱὸν δ' οὐ τέκετ' ἄλλον ἐπὶ κτεάτεσσι λιπέσθαι.
ἔνθ' ὅ γε τοὺς ἐνάριζε, φίλον δ' ἐξαίνυτο θυμὸν    155
ἀμφοτέρω, πατέρι δὲ γόον καὶ κήδεα λυγρὰ
λεῖπ', ἐπεὶ οὐ ζώοντε μάχης ἐκνοστήσαντε
δέξατο· χηρωσταὶ δὲ διὰ κτῆσιν δατέοντο.
Ἔνθ' υἷας Πριάμοιο δύω λάβε Δαρδανίδαο
εἰν ἑνὶ δίφρωι ἐόντας Ἐχέμμονά τε Χρομίον τε.    160
ὡς δὲ λέων ἐν βουσὶ θορὼν ἐξ αὐχένα ἄξηι
πόρτιος ἠὲ βοὸς ξύλοχον κάτα βοσκομενάων,
ὣς τοὺς ἀμφοτέρους ἐξ ἵππων Τυδέος υἱὸς
βῆσε κακῶς ἀέκοντας, ἔπειτα δὲ τεύχε' ἐσύλα·
ἵππους δ' οἷς ἑτάροισι δίδου μετὰ νῆας ἐλαύνειν.    165

In the first simile the lion plunders a sheepfold. The shepherd only manages to graze the attacker and now cowers in the steading. The lion's assault reduces the sheep to mere heaps. The lion then leaps out of the enclosure but still rages. We are told that the shot only stirred up the strength in the lion's breast; implicit is the threat that he will return to avenge himself with renewed vigor. The simile serves to reintegrate Diomedes into the fighting with amplified stature. The repetition of a form of μίγνυμι (ἐμίχθη, 134) in the resumptive clause (μίγη, 143) emphasizes this reintegration. The first lion simile is pivotal to the action in the narrative. Just as it refers back to the preceding narrative, so it looks ahead.

The use of heroic and emotive vocabulary in this passage is striking. μεμαώς is repeated three times—once in the verse before the simile (135), once in the simile proper (142), and once in the resumptive clause (143). Athene's inspiration of μένος (a cognate of μεμαώς) in the warrior (125, cf. 136) is mirrored in the simile by τοῦ μέν τε σθένος ὦρσεν (139). In condensed form, the simile recapitulates the preceding narrative events. Like Pandaros' ineffectual arrow the spear–shot of the shepherd scarcely grazes the lion. οὐδὲ δαμάσσῃ (138) echoes τὸν δ' οὐ βέλος ὠκὺ δάμασσεν (106). The spear–shot only serves to stir up the lion's strength, reflecting Diomedes' piqued pride and the augmentation of his strength through Athene. This detail is echoed later in Pandaros' complaint to Aineias that his arrows only awakened the anger in his opponents, Diomedes and Menelaos, all the more: ἤγειρα δὲ μᾶλλον (208).

The simile is open–ended. The wounded lion of the simile virtually jumps out of the yard to avenge himself in the next developed simile, which describes a lion killing a calf or an ox (161ff.). Even though the two similes are separated by sixteen lines of narrative, and the object of the lion's attack alters from sheep to cows, they are clearly meant to be taken together sequentially as a parallel of the narrative situation. Internally the second simile resolves the action of the first. That it is meant to be a continuation of the first is indicated by the fact that no introductory verb or emotive term is required to launch it: it simply begins at the beginning of the verse. This simile, in contrast to the previous one, emphasizes action.

The lion in the second simile jumps on its victim's neck, which accurately portrays the lion's method of attack: the momentum of the leap makes the prey fall and break its neck under the weight.[3] Once the victim is down the lioness sinks her teeth into the throat, a formulaic element of lion similes in *aristeiai*.[4] Although the second simile follows the account of the first of three pairs of slain warriors, it sums up the killing of all three.

In particular the details of the wounding of Hypeiron, one of the first pair, blend with the simile. Diomedes' sword passes through the collar–bone and hacks the neck and back from the shoulder of his victim, Hypeiron, called "Shepherd of the people." The trauma to the neck (αὐχένος, 147) of the human victim modulates in the simile to the broken neck (αὐχένα, 161) of the animal victim. The metaphorical usage of ποιμήν in the epithet ποιμένα λαῶν (144) interacts with the literal use of ποιμήν in the simile.[5] Diomedes' slaying of Hypeiron carries out in the narrative the implied revenge of the lion. Hypeiron, "shepherd of the people," thus corresponds to both the shepherd and the sheep in the simile. As revenge for

---

[3] H. W. Auden, *CR* 10 (1896) 107; cf. G. Guggisberg, *Simba, the Life of the Lion* (1961) 122.

[4] Apart from the present example, it occurs in Agamemnon's and Menelaos' *aristeiai* (11.175=17.63).

[5] The name Hypeiron coincides with the frequent epithet and synonym of Helios, who is renowned for his herds of cattle. Hyp– names in the *Iliad* are generic and used with the formula ποιμένι or -α λαῶν to designate minor warriors in confrontation with a major warrior (5.144, 7.469, 11.578, 11.92, 13.411, 14.516.) The use of the honorific epithet at the moment of confrontation makes the bearer of the title seem a more worthy opponent for his attacker but also introduces a note of pathos, like the condensed biographies accompanying the deaths of minor warriors.

being wounded in the upper body (shoulder), Diomedes slays his victim in the manner of the avenging lion who attacks his victim in the upper body (neck) in the ensuing simile.[6] The analogy between the warrior's neck wound and the lion's assault on the neck is part of a pattern.[7] In the narrative preceding a simile in Menelaos' *aristeia* the warrior had driven the spear through the soft part of Euphorbos' neck (17.49f.). Likewise, in the narrative following the simile in Agamemnon's *aristeia* some of the human victims are said to fall "flat on their faces," others "on their backs" (11. 179), much as one would expect the animal victims to fall under the weight of the lion. A similar situation occurs later in the *aristeia*. At 11.239 Agamemnon is compared in a short simile to a lion. In the next lines his victim Iphidamas, a man who ironically gave large ἔεδνα of cattle, falls from a wound in the neck. Although wounds in the neck are common enough (six times in the *Iliad*), these three examples probably reflect the lion's method of attack.

> Ἔνθ' αὖτ' Αἰνείας Δαναῶν ἕλεν ἄνδρας ἀρίστους      5.541
> υἷε Διοκλῆος Κρήθωνά τε 'Ορσίλοχόν τε,
> τῶν ῥα πατὴρ μὲν ἔναιεν ἐϋκτιμένηι ἐνὶ Φηρῆι
> ἀφνειὸς βιότοιο, γένος δ' ἦν ἐκ ποταμοῖο
> 'Αλφειοῦ, ὅς τ' εὐρὺ ῥέει Πυλίων διὰ γαίης,      545
> ὃς τέκετ' 'Ορτίλοχον πολέεσσ' ἄνδρεσσιν ἄνακτα·
> 'Ορτίλοχος δ' ἄρ' ἔτικτε Διοκλῆα μεγάθυμον,
> ἐκ δὲ Διοκλῆος διδυμάονε παῖδε γενέσθην,
> Κρήθων 'Ορσίλοχός τε μάχης εὖ εἰδότε πάσης.
> τὼ μὲν ἄρ' ἡβήσαντε μελαινάων ἐπὶ νηῶν      550
> Ἴλιον εἰς εὔπωλον ἅμ' 'Αργείοισιν ἐπέσθην,
> τιμὴν 'Ατρείδηις 'Αγαμέμνονι καὶ Μενελάωι
> ἀρνυμένω· τὼ δ' αὖθι τέλος θανάτοιο κάλυψεν.
> οἵω τώ γε λέοντε δύω ὄρεος κορυφῆισιν
> ἐτραφέτην ὑπὸ μητρὶ βαθείης τάρφεσιν ὕλης·      555
> τὼ μὲν ἄρ' ἁρπάζοντε βόας καὶ ἴφια μῆλα
> σταθμοὺς ἀνθρώπων κεραΐζετον, ὄφρα καὶ αὐτὼ
> ἀνδρῶν ἐν παλάμηισι κατέκταθεν ὀξέι χαλκῶι·
> τοίω τὼ χείρεσσιν ὑπ' Αἰνείαο δαμέντε.

---

[6] Pandaros, of course, is the man on whom Diomedes wants to gain vengeance, and he does so later in the Book. His first victims are, as it were, scapegoats.

[7] Cf. M. Willcock, *A Companion to Homer* (1976) 280, noting that warriors who fall "like a tree" have been hit in the head; G. Strasburger, *Die kleinen Kämpfer der Ilias* (diss. Frankfurt 1954) 38f.

καππεσέτην, ἐλάτηισιν ἐοικότες ὑψηλῆισι.                560

The third developed lion simile (5.554ff.) provides an unusually fine example of the blending of internal pictorial content and external narrative detail. Here, as we will see in the *aristeia* of Agamemnon, simile mirrors biography. The use of the dual throughout reinforces the impression that the simile was carefully adapted to suit the narrative context. The simile follows closely upon Sarpedon's rebuke to Hektor in which he characterizes the latter's troops as cowering like hounds around a lion (476). Ares intervenes to incite the Trojans, while the Aiantes, Odysseus, and Diomedes stir the Danaans. In the ensuing battle the major warriors on both sides encounter an opponent. Among the Trojans Aineias kills two Danaans, Krethon and Orsilochos, twins born to Diokles of Phere. A condensed biography charts their lives. They were skilled in every way in fighting; when they reached manhood they joined the common cause of the Achaians and set sail in the black ships to win honor for Agamemnon and Menelaos. But now, unfortunately, "the fulfillment of death was a darkness upon them."

The simile in complementary fashion traces the life–cycle of a pair of lion cubs. Their mother nourishes them in the deep forest. When they are grown they plunder the cows and sheep in the steadings of men until one day they are struck with the sharp bronze. The simile concludes with a coda comparing the twins in a short simile to two tall pines, ἐλάτηισιν ἐοικότες ὑψηλῆισι (560). The tree simile in fact should be taken together with the lion simile as an organic unit, since it corresponds with the forest setting (ὄρεος κορυφῆισιν ... ἐτραφέτην ὑπὸ μητρὶ βαθείης τάρφεσιν ὕλης) in the beginning verse of the lion simile (554f.).[8] The mountainous setting also harmonizes with the name given for the twin's place of birth in Phere.[9] The pathetic description of the lives of the cubs provides an epitaph in the animal metaphor for the condensed biography of the warriors on the battlefield. Both report birth and parentage and coming of age, participation in warfare, and finally, premature death. The parallel biographies wed the rhythms of life in human and animal spheres. Man and animal are seen to be yoked to the same uncompromising fates.

---

[8] G. Strasburger, *op. cit.* 37f., discusses the lion and tree simile as a variation on tree similes (4.482, 17.53, cf. 8.306) that compare the life and death of a minor warrior to the stages of growth of trees before they are destroyed by a human or natural force.

[9] Cf. Φῆρές 1.268, 2.743, and θήρ. Phere is represented as being half–way between Pylos and Sparta (*Od.* 3.488).

## The *Aristeia* of Agamemnon: Book 11

Apart from one short wolf simile,[10] animal imagery in the *aristeia* of the king and chief commander of the Achaians is restricted to the lion. The latent regal associations of the lion are strongly brought out in this episode. In addition to the two lion similes to be considered, two short lion similes occur in the *aristeia* at 129 and 239. The lion similes complement the description of the snake on Agamemnon's telamon (38–40) in the arming scene. On it writhes a three–headed snake—one of the four fearless animals of Proteus' transformation. Also depicted on the telamon is Φόβος, which will recur in the similes in verbal forms as an example of the overlap between simile and ecphrasis. The lion in the developed similes may be seen as the active manifestation of the force embodied in the static device on the telamon. The lion simile complements a fire simile at 155ff.; the phrase αἰὲν ἀποκτείνων embraces the pair (154, 178). The simile of fire obliterating trees in a forest is a violent culmination of two earlier plant similes picturing reapers (67ff.) and wood–cutters (86ff.).

Ἀτρεΐδης δ' ἕπετο σφεδανὸν Δαναοῖσι κελεύων.    11.165
οἳ δὲ παρ' Ἴλου σῆμα παλαιοῦ Δαρδανίδαο
μέσσον κὰπ πεδίον παρ' ἐρινεὸν ἐσσεύοντο
ἱέμενοι πόλιος· ὃ δὲ κεκλήγων ἕπετ' αἰεὶ
Ἀτρεΐδης, λύθρωι δὲ παλάσσετο χεῖρας ἀάπτους.
ἀλλ' ὅτε δὴ Σκαιάς τε πύλας καὶ φηγὸν ἵκοντο,    170
ἔνθ' ἄρα δὴ ἵσταντο καὶ ἀλλήλους ἀνέμιμνον.
οἳ δ' ἔτι κὰμ μέσσον πεδίον φοβέοντο βόες ὥς,
ἅς τε λέων ἐφόβησε μολὼν ἐν νυκτὸς ἀμολγῶι
πάσας· τῆι δέ τ' ἰῆι ἀναφαίνεται αἰπὺς ὄλεθρος·
τῆς δ' ἐξ αὐχέν' ἔαξε λαβὼν κρατεροῖσιν ὀδοῦσι    175
πρῶτον, ἔπειτα δέ θ' αἷμα καὶ ἔγκατα πάντα λαφύσσει·
ὣς τοὺς Ἀτρεΐδης ἔφεπε κρείων Ἀγαμέμνων
αἰὲν ἀποκτείνων τὸν ὀπίστατον· οἳ δ' ἐφέβοντο.
πολλοὶ δὲ πρηνεῖς τε καὶ ὕπτιοι ἔκπεσον ἵππων
Ἀτρεΐδεω ὑπὸ χερσί· περὶ πρὸ γὰρ ἔγχεϊ θῦεν.    180
ἀλλ' ὅτε δὴ τάχ' ἔμελλεν ὑπὸ πτόλιν αἰπύ τε τεῖχος

---

[10] 11.72: used not of Agamemnon but to emphasize close fighting among the ranks of men. Moulton, *Similes* 96–99, discusses the use of the similes in Book 11 to characterize Agamemnon; cf. Schnapp–Gourbeillon, *Lions* 72–73.

ἵξεσθαι, τότε δή ῥα πατὴρ ἀνδρῶν τε θεῶν τε
Ἴδης ἐν κορυφῆισι καθέζετο πιδηέσσης
οὐρανόθεν καταβάς· ἔχε δ' ἀστεροπὴν μετὰ χερσίν.
Ἶριν δ' ὄτρυνε χρυσόπτερον ἀγγελέουσαν·    185

The simile occurs at the heart of the *aristeia* of Agamemnon. At this point in the episode the chief commander succeeds in routing two groups of Trojans, one of which cannot run fast enough to make a new stand with others at the Skaian gates. They are left straggling in the plain and become, one by one, Agamemnon's victims. Agamemnon's triumph occasions the intervention of Zeus, who dispatches Iris to Hektor. Internally the simile pictures a lion stampeding a herd of cattle (φοβέοντο 172). Like a robber the beast stalks his prey at night, μολὼν ἐν νυκτὸς ἀμολγῷ, he frightens the entire herd ἐφόβησε … πάσας. Note the repetition of φοβέω in ἐφόβησε. The next couplet contains two careful contrasts marked by the adversative δέ in 174: first, between the entire herd and one victim πάσας)(τῇ δέ τ' ἰῇ and second, between dark and light in νυκτὸς ἀμολγῷ)(ἀναφαίνεται. An example of pronounced parataxis occurs in the following two couplets which describe first (πρῶτον) the lethal attack of the strong–toothed lion on his prey and then (ἔπειτα) the feast.

Although the simile proper begins at verse 172 the lion subject is anticipated three lines earlier in the narrative by figurative language found elsewhere in lion similes. Agamemnon's bloody appearance is stressed by the phrase λύθρῳ παλάσσετο χεῖρας ἀάπτους (169). The formula in modified form recurs in two lion similes that emphasize how Odysseus looked after killing the suitors: αἵματι καὶ λύθρῳ πεπαλαγμένον ὥς τε λέοντα; one of the similes develops the image of the gory lion (αἱματόεντα, *Od.* 22.405) as he feasts on a cow.[11] The simile in Agamemnon's *aristeia* also goes on to report the lion's bloody feast (sc. αἷμα 176). The simile is launched by the phrase βόες ὥς, preceded by the verb φοβέοντο. It begins from the perspective of the victims, not the lion. The locomotive and emotive functions of the simile can be understood through the triple repetition of φοβέω. In the *Iliad* φοβέω is more often than not an action verb meaning "to put to flight" in the active voice, and in the passive voice, "to flee." Thus it expresses movement; but it also has a strong emotive undercurrent expressing the attendant fear. Both components are present in the simile. From the point of view of the action the simile brings to an end the

---

[11] *Od.* 22.402=*Od.* 23.48; the phrase is not restricted to lion similes.

mass attack of the chief commander. Agamemnon pursues the stragglers; φοβέοντο and ἐφέβοντο mark this collective flight. In emotive terms the repeated use of φοβέω emphasizes the frightening aspect of the aggressor (earlier personified as Φόβος on the telamon) and of the dread of the victims. We are made to see the attack both from the point of view of the aggressor and the victim, as the use of the active and passive voices of φοβέω implies. To Agamemnon, as to the lion, the best victim is that which falls easiest prey—the one which straggles behind. In the resumptive clause of the simile Agamemnon is said "always to kill the last one," αἰὲν ἀποκτείνων τὸν ὀπίστατον. From the victims' point of view we experience both the collective and individual fear of the victims. The stampeding herd corresponds to the mass flight of the Trojans, while the fate of the lion's chosen victim expresses the fear which seizes upon each man in the random and indiscriminate attack. At this climactic moment in the episode the simile gives an insight into the fear preying upon the individual.

αὐτὰρ ὃ βῆ Ἰσόν τε καὶ Ἄντιφον ἐξεναρίξων    101
υἷε δύω Πριάμοιο νόθον καὶ γνήσιον ἄμφω
εἰν ἑνὶ δίφρωι ἐόντας· ὃ μὲν νόθος ἡνιόχευεν,
Ἄντιφος αὖ παρέβασκε περικλυτός· ὥ ποτ' Ἀχιλλεὺς
Ἴδης ἐν κνημοῖσι δίδη μόσχοισι λύγοισι,    105
ποιμαίνοντ' ἐπ' ὄεσσι λαβών, καὶ ἔλυσεν ἀποίνων.
δὴ τότε γ' Ἀτρείδης εὐρὺ κρείων Ἀγαμέμνων
τὸν μὲν ὑπὲρ μαζοῖο κατὰ στῆθος βάλε δουρί,
Ἄντιφον αὖ παρὰ οὖς ἔλασε ξίφει, ἐκ δ' ἔβαλ' ἵππων.
σπερχόμενος δ' ἀπὸ τοῖιν ἐσύλα τεύχεα καλὰ    110
γιγνώσκων· καὶ γάρ σφε πάρος παρὰ νηυσὶ θοῆισιν
εἶδεν, ὅτ' ἐξ Ἴδης ἄγαγεν πόδας ὠκὺς Ἀχιλλεύς.
ὡς δὲ λέων ἐλάφοιο ταχείης νήπια τέκνα
ῥηϊδίως συνέαξε λαβὼν κρατεροῖσιν ὀδοῦσιν
ἐλθὼν εἰς εὐνήν, ἁπαλόν τέ σφ' ἦτορ ἀπηύρα·    115
ἣ δ' εἴ πέρ τε τύχηισι μάλα σχεδόν, οὐ δύναταί σφι
χραισμεῖν· αὐτὴν γάρ μιν ὑπὸ τρόμος αἰνὸς ἱκάνει·
καρπαλίμως δ' ἤιξε διὰ δρυμὰ πυκνὰ καὶ ὕλην
σπεύδουσ' ἱδρώουσα κραταιοῦ θηρὸς ὑφ' ὁρμῆς·
ὡς ἄρα τοῖς οὔ τις δύνατο χραισμῆσαι ὄλεθρον    120
Τρώων, ἀλλὰ καὶ αὐτοὶ ὑπ' Ἀργείοισι φέβοντο.

The poet's sympathy for the victim can be observed in a developed lion simile earlier in Agamemnon's *aristeia*. The simile is the companion to a condensed biography of Isos and Antiphos, the second of three pairs of victims in a brief catalogue (91–147). The simile is not the marauding type, but rather a variation describing a lion invading the lair of a doe. The beast devours the young while the mother looks on helpless; fearing for her own life, she bursts into flight. In this developed simile of seven lines, four are devoted to the nightmarish reaction of the doe. The anthropomorphic vocabulary used for the fawns increases the pathos we feel for the mother. Her young are called νήπια τέκνα, and the heart of the victim is said to be ἀπαλόν, "delicate." "So," we are told in the resumptive clause, "there was no one of the Trojans who could save the two from death..."

The technique of making the sympathy of the audience side with the victim closely parallels the attention drawn to minor warriors through biographies. Here the lion–doe simile complements the earlier tale of the lives of Isos and Antiphos. It relates how earlier Achilles apprehended the sons of Priam, one a bastard, and the other legitimate, while they were watching their sheep in the folds of Mt. Ida. Achilles released them, however, for a ransom. Now on the battlefield Agamemnon kills them off. He spears Isos in the chest above the nipple and hurls Antiphos from the chariot with his sword. Eagerly the chief commander rushes in and strips off the armor of the pair he knew, for "he had seen these two before by the fast ships | When Achilles of the swift feet had brought them in from Ida."

The simile parallels the biography and killing both in setting and in minor details. The invasion of the doe's lair resembles Achilles' disruption of the herdsmen in a sequestered area, in the folds of Mt. Ida (Ἴδης ἐν κνημοῖσι, 105). The lion prises the heart from his innocent victims and Agamemnon spears the bastard Isos in the chest above the nipple. The verb for ripping out the soft heart, ἀπηύρα (115), used elsewhere of forcibly depriving one of something valuable—including a corpse of its armor (11.334, 432, 17.125)—here echoes the act of despoiling in line 110. The theme of the mother bereft of her young is explicit in the simile and suggested in the biography. In tending their sheep Isos and Antiphos perform a function akin to child–rearing. Achilles' disturbance of pastoral life prevents the pair from carrying out their herding duties, as it deprives them of their biological function as fathers.

The conclusion of this simile demonstrates how similes sometime serve to resume the tempo and redirect the narrative. The biography and subsequent simile have introduced a *ritardando* into the catalogue.   But then in the resumptive clause of the simile the doe's flight is compared to the Trojans' flight in general.  The doe functions as a medium for returning thesimile to the larger narrative scene at hand. The poet manipulates the emotions of the audience by making their sympathy linger over the plight of the doe; suddenly the emphasis modulates from emotion for the individual animal to the mass motion of the throng. The slaying continues. The chief commander is compared in a short simile to a lion (129); his victims supplicate, promising a ransom (131ff.) which he denies with characteristic cruelty. The shorter simile and ransom scene echo the biography and the developed lion simile (113ff.) considered above.

## Reversal: Hektor Storms the Wall: Book 12

The Achaians are the aggressors against Troy, and their military objective is to penetrate the Trojan citadel and capture the possessions within. In the *Teichomachia*, where the Trojans gain the upper hand in the fighting, the tactical situation is reversed. Hektor and Sarpedon lead the Trojans in an attack against the Achaian wall, which can be likened to a temporary enceinte wall. In marauding lion similes, as we have seen, the lion's goal is to penetrate the cattle steading, and thus it resembles the basic military initiative.

There are three major attacks on the wall in Book 12. The first, attempted by the foolhardy Asios, is quickly thwarted by Leonteus and Polypoites. The second attack, carried out by Sarpedon, is somewhat more successful (330ff.), and eventually he manages to open a breach in the wall (397–99). Two marauding lion similes, one short and one developed, precede his attack (293, 299ff.). Sarpedon's efforts facilitate the successful storming of the gates by Hektor. No simile marks his triumphant entry. The first simile in Book 12, however, compares Hektor, as he tries to cross the ditch fortified by palisades, to a lion or boar wheeling on men and dogs "closing themselves into a wall about him."

It is a curious feature of the lion imagery that Hektor (and once in Book 12, Sarpedon) is the only individual Trojan accorded marauding lion similes, and that they complement his attack on the Achaian wall or ships.[12] The lion is more often than not an emblem of the Achaians, who are the aggressors desirous of overleaping the walls of the citadel. The shift in the object of comparison in Books 12 and 15 corresponds to the temporary reversal of the tactical situation. In the discussion that follows the similes will be considered individually in order to discover how they respond to the requirements of the narrative.

Ἕκτορα δειδιότες, κρατερὸν μήστωρα φόβοιο·          12.39
αὐτὰρ ὅ γ' ὡς τὸ πρόσθεν ἐμάρνατο ἶσος ἀέλληι·
ὡς δ' ὅτ' ἂν ἔν τε κύνεσσι καὶ ἀνδράσι θηρευτῆισι
κάπριος ἠὲ λέων στρέφεται σθένεϊ βλεμεαίνων·
οἳ δέ τε πυργηδὸν σφέας αὐτοὺς ἀρτύναντες
ἀντίον ἵστανται καὶ ἀκοντίζουσι θαμειὰς
αἰχμὰς ἐκ χειρῶν· τοῦ δ' οὔ ποτε κυδάλιμον κῆρ          45
ταρβεῖ οὐδὲ φοβεῖται, ἀγηνορίη δέ μιν ἔκτα·
ταρφέα τε στρέφεται στίχας ἀνδρῶν πειρητίζων·
ὅππηι τ' ἰθύσηι τῆι εἴκουσι στίχες ἀνδρῶν·
ὡς Ἕκτωρ ἀν' ὅμιλον ἰὼν εἰλίσσεθ' ἑταίρους

The first simile, comparing Hektor to a boar or lion,[13] is linked with the third and developed marauding lion simile by shared vocabulary: ἀ-κοντίζουσι ... χειρῶν (44f.), cf. χειρὸς ἄκοντι (306); ἀγηνορίη (46) cf. θυμὸς ἀγήνωρ (300); πειρητίζων (47), cf. πειρήσοντα, ἀπείρητος (301, 304). Internally this first simile describes a beast who turns at bay (στρέφεται) on his attackers. The men and dogs make a ring or wall (πυργηδόν) about the beast and and then proceed to throw spears at him.

---

[12] A marauding lion simile occurs at 18.161 as Hektor fights over Patroklos' corpse. Hektor and Aias are compared to lions and boars fighting (7.256f.); Euphorbos' pride is said to be more excessive than that of a lion, leopard, or boar (17.20ff.). Aineias is compared to a lion in a short lion simile at 5.299. Hektor is compared to a lion when confronting Patroklos in the famous lion–boar simile at 16.823ff., cf. 756ff. (Hektor and Patroklos). The Trojans are briefly equated with lions in a short simile at 15.592.

[13] The suitability of this simile to the narrative has been the subject of discussion since antiquity. The simile properly belongs to a tactical situation in which a warrior is encircled, sc. Odysseus at 11.412ff., followed by a simile of a boar hemmed in his lair. Such a simile contains a trace of the suggestion that the poet drew his simile from a store of types, some of which were more appropriately bridged with the narrative than others. Moulton, *Similes* 47f., n. 54, summarizes the views and rightly places greater emphasis on the psychological correspondence between the simile and narrative than on parallels in the action. J. Latacz, *Kampfparänese, Kampfdarstellung und Kampfwirklichkeit der Ilias, bei Kallinos und Tyrtaios* (1977) 57, cf. 165, includes this simile among those which describe the phalanx led in an offensive maneuver.

The courageous animal feels no terror in his heart, nor flees, "and it is his own courage that kills him." (46) The simile goes on to describe the beast repeatedly (ταρφέα) trying to break the ranks of men (στίχας ἀνδρῶν), a phrase recurring in the next verse (στίχες ἀνδρῶν). In the resumptive clause the stalwart warrior is described as wheeling among the throng (εἰλίσσεθ').[14]

There is a juggling of tactical and animal vocabulary between simile and narrative. The verb in the resumptive clause, εἰλίσσεθ', elsewhere describes the movement of the boar at bay (17.283, 728). The repetition of the phrase στίχες (–ας) ἀνδρῶν lends a tactical character to the couplet which continues in the simile after the proleptic mention of the animal's death. ταρφέα (a *hapax legomenon* adverbially) evokes the haunts of the lion, who in a simile is said to be raised in the deep forest, βαθείης τάρφεσιν ὕλης (5.555, cf. 15.606). The most pronounced tactical usage in the simile is πυργηδόν, "in a wall" (cf. 15.618).[15] It likens the human subjects in the simile to a wall, which is the inanimate leitmotif of the *Teichomachia*. Its use in Book 12 precisely reflects the narrative situation of Trojans in battle with Achaians throwing spears from the ramparts. πύργος, the word for "rampart" occurs ten times in this Book. The lion–boar simile marks Hektor's attack not on the wall but rather his efforts to cross the ditch protected by palisades.

The developed lion simile at 12.299ff. has significant connections with its context, which includes a short lion simile, an arming scene, and a speech.[16] The short lion simile at 293 reannounces the animal metaphor. It occurs in the protasis of a past contrary–to–fact conditional which states that Hektor never would have broken the wall and the large bolt (τείχεος ἐρρήξαντο πύλας καὶ μακρὸν ὀχῆα, 291) had Zeus not sent Sarpedon against the Argives as a lion stampedes the cattle. The more elaborate lion simile, beginning six verses later, can be seen as an intensification of the shorter lion simile.[17] Together the two similes build up the momentum of Sarpedon's imminent attack. Sarpedon's well–known *parainesis* to Glaukos, however, intervenes to retard the action. Its function is similar to the suspenseful flash–back to the Parnassos boar hunt just after Eurykleia has

---

[14] The OCT follows the vulgate reading ἐλ(λ)ίσσεθ', "entreated."

[15] Cf. 15.618.

[16] Cf. Krischer, *Formale Konventionen* 40, "*Ansturm zum Kampf.*"

[17] Moulton, *Similes* 20–21, sees the pairing of a short and developed simile as an example of *auxesis*.

recognized Odysseus by his wound.[18] The actual attack is held off until the end of the speech (329ff.).

Οὐδ’ ἄν πω τότε γε Τρῶες καὶ φαίδιμος Ἕκτωρ          12.290
τείχεος ἐρρήξαντο πύλας καὶ μακρὸν ὀχῆα,
εἰ μὴ ἄρ’ υἱὸν ἐὸν Σαρπηδόνα μητίετα Ζεὺς
ὦρσεν ἐπ’ Ἀργείοισι λέονθ’ ὣς βουσὶν ἕλιξιν.
αὐτίκα δ’ ἀσπίδα μὲν πρόσθ’ ἔσχετο πάντοσ’ ἐΐσην
καλὴν χαλκείην ἐξήλατον, ἣν ἄρα χαλκεὺς          295
ἤλασεν, ἔντοσθεν δὲ βοείας ῥάψε θαμειὰς
χρυσείῃς ῥάβδοισι διηνεκέσιν περὶ κύκλον.
τὴν ἄρ’ ὅ γε πρόσθε σχόμενος δύο δοῦρε τινάσσων,
βῆ ῥ’ ἴμεν ὥς τε λέων ὀρεσίτροφος, ὅς τ’ ἐπιδευὴς
δηρὸν ἔῃ κρειῶν, κέλεται δέ ἑ θυμὸς ἀγήνωρ          300
μήλων πειρήσοντα καὶ ἐς πυκινὸν δόμον ἐλθεῖν·
εἴ περ γάρ χ’ εὕρῃσι παρ’ αὐτόφι βώτορας ἄνδρας
σὺν κυσὶ καὶ δούρεσσι φυλάσσοντας περὶ μῆλα,
οὔ ῥά τ’ ἀπείρητος μέμονε σταθμοῖο δίεσθαι,
ἀλλ’ ὅ γ’ ἄρ’ ἢ ἥρπαξε μετάλμενος, ἠὲ καὶ αὐτὸς          305
ἔβλητ’ ἐν πρώτοισι θοῆς ἀπὸ χειρὸς ἄκοντι·
ὣς ῥα τότ’ ἀντίθεον Σαρπηδόνα θυμὸς ἀνῆκε
τεῖχος ἐπαΐξαι διά τε ῥήξασθαι ἐπάλξεις.
αὐτίκα δὲ Γλαῦκον προσέφη παῖδ’ Ἱππολόχοιο·
Γλαῦκε τί ἦ δὴ νῶι τετιμήμεσθα μάλιστα          310
ἕδρῃ τε κρέασίν τε ἰδὲ πλείοις δεπάεσσιν
ἐν Λυκίῃ, πάντες δὲ θεοὺς ὣς εἰσορόωσι,
καὶ τέμενος νεμόμεσθα μέγα Ξάνθοιο παρ’ ὄχθας
καλὸν φυταλιῆς καὶ ἀρούρης πυροφόροιο;
τῶ νῦν χρὴ Λυκίοισι μέτα πρώτοισιν ἐόντας          315
ἑστάμεν ἠδὲ μάχης καυστείρης ἀντιβολῆσαι,
ὄφρά τις ὧδ’ εἴπῃ Λυκίων πύκα θωρηκτάων·
οὐ μὰν ἀκλεέες Λυκίην κάτα κοιρανέουσιν
ἡμέτεροι βασιλῆες, ἔδουσί τε πίονα μῆλα
οἶνόν τ’ ἔξαιτον μελιηδέα· ἀλλ’ ἄρα καὶ ἲς          320
ἐσθλή, ἐπεὶ Λυκίοισι μέτα πρώτοισι μάχονται.
ὦ πέπον εἰ μὲν γὰρ πόλεμον περὶ τόνδε φυγόντε
αἰεὶ δὴ μέλλοιμεν ἀγήρω τ’ ἀθανάτω τε
ἔσσεσθ’, οὔτέ κεν αὐτὸς ἐνὶ πρώτοισι μαχοίμην

---

[18] *Od.* 19.392–466

οὔτέ κε σὲ στέλλοιμι μάχην ἐς κυδιάνειραν·      325
νῦν δ' ἔμπης γὰρ κῆρες ἐφεστᾶσιν θανάτοιο
μυρίαι, ἃς οὐκ ἔστι φυγεῖν βροτὸν οὐδ' ὑπαλύξαι,
ἴομεν ἠέ τωι εὖχος ὀρέξομεν ἠέ τις ἡμῖν.

The pictorial content of the simile describes a lion urgent to enter the sheep–stead and to make trial of (πειρήσοντα) the shepherd's hut, called a close–set "house" (δόμον). πειρητίζων (47), one of the words which links the lion–boar simile with the developed lion simile (sc. πειρήσοντα, 301) recurs in battle narrative at 257, where the Trojans, under the leadership of Hektor, try to break down the walls and tear away the ramparts (257f.): ῥήγνυσθαι μέγα τεῖχος 'Αχαιῶν πειρήτιζον. Ι κρόσσας μὲν πύργων ἔρυον... Here, what was faintly perceptible in Hektor's lion–boar simile, is now explicit: the staging of the simile closely matches the narrative situation of Sarpedon attempting to penetrate the Achaian wall. At 290ff. it is stated that Zeus sends his son to help Hektor and the Trojans break the gates of the rampart; Sarpedon's own desire to storm the gates is made known in the resumptive clause of the simile. Pictorially the simile conveys what Sarpedon verbalizes. Through the animal metaphor it states: "this is the fate of the courageous warrior risking attack." The concluding two lines of the lion simile (12.305f.):

ἀλλ' ὅ γ' ἄρ' ἢ ἥρπαξε μετάλμενος, ἠὲ καὶ αὐτὸς
ἔβλητ' ἐν πρώτοισι θοῆς ἀπὸ χειρὸς ἄκοντι·

recall the either ... or construction in the concluding lines of Sarpedon's speech marking his resolve to fight (327f.):

...ἃς οὐκ ἔστι φυγεῖν βροτὸν οὐδ' ὑπαλύξαι,
ἴομεν ἠέ τῳ εὖχος ὀρέξομεν, ἠέ τις ἡμῖν.

The simile concentrates on exploring the motives and intentions of the lion. He is a lion raised in the mountains. The lion attacks the sheep–steading because he has gone for a long time without food: ὅς τ' ἐπιδευὴς Ι δηρὸν ἔῃ κρειῶν (299f.); then in the second half of the line the psychological motive for his attack is given: κέλεται δέ ἑ θυμὸς ἀγήνωρ "his manly spirit bid him." As the simile develops it continues to probe further into the mind of the lion, who, despite the odds, is not driven from the steading without making an attack (οὐ ῥά τ' ἀπείρητος μέμονε, 304), and either springs to seize a sheep or else is hit in the first skirmish. The passage does not tell us the actual outcome of the lion's onslaught; rather it ends on a speculative note that heightens the suspense of the present situation.

Sarpedon's *parainesis* to Glaukos restates in human terms certain details from the developed simile as the King of Lykia reminds his second in command of the duties incumbent upon the king in battle. The speech characterizes a kind of feudal relationship between the kings of Lykia and their people. He risks his life fighting for his people in the front ranks, and in return is rewarded with the choice pieces of meat: the flesh of fat sheep and sweet wine. In order to uphold his contract Sarpedon must never quail. For the king, as for the lion, action is the only recourse; he must risk attack in the hope of winning fame.

The *parainesis* mirrors the appetites motivating the lion. There is a parallel between the king's reward and the lion's, supported by two linguistic correspondences. The beast's recompense lies in the sheep he may devour, the king's in the flesh of fat sheep served to him in time of peace. κρειῶν in the phrase δηρὸν ἔη κρειῶν (300) corresponds to κρέασιν (311), and μῆλα (303) to μῆλα (319). The anthropomorphic motivation attributed to the lion, called a bidding θυμὸς ἀγήνωρ, is answered in Sarpedon's *speech by εὖχος (328). For the warrior, the desire for glory is as much a* motivating appetite as animal flesh is for the lion. Neither beast nor man will retreat before the opportunity of sating these appetites. The creation of a lion with a manly spirit, a reasoning mind, and indomitable courage—in short a kind of hybrid lion—emphasizes that both bestial and human appetites spur on the warrior.

### Hektor Storms the Ships: Book 15

Ἀργεῖοι δ' ὑπέμειναν ἀολλέες, ὦρτο δ' αὐτὴ          15.312
ὀξεῖ' ἀμφοτέρωθεν, ἀπὸ νευρῆφι δ' ὀιστοὶ
θρῶισκον· πολλὰ δὲ δοῦρα θρασειάων ἀπὸ χειρῶν
ἄλλα μὲν ἐν χροὶ πήγνυτ' ἀρηϊθόων αἰζηῶν,          315
πολλὰ δὲ καὶ μεσσηγὺ πάρος χρόα λευκὸν ἐπαυρεῖν
ἐν γαίηι ἵσταντο λιλαιόμενα χροὸς ἆσαι.
ὄφρα μὲν αἰγίδα χερσὶν ἔχ' ἀτρέμα Φοῖβος Ἀπόλλων,
τόφρα μάλ' ἀμφοτέρων βέλε' ἥπτετο, πῖπτε δὲ λαός.
αὐτὰρ ἐπεὶ κατ' ἐνῶπα ἰδὼν Δαναῶν ταχυπώλων          320
σεῖσ', ἐπὶ δ' αὐτὸς ἄυσε μάλα μέγα, τοῖσι δὲ θυμὸν
ἐν στήθεσσιν ἔθελξε, λάθοντο δὲ θούριδος ἀλκῆς.
οἳ δ' ὥς τ' ἠὲ βοῶν ἀγέλην ἢ πῶυ μέγ' οἰῶν

θῆρε δύω κλονέωσι μελαίνης νυκτὸς ἀμολγῶι
ἐλθόντ' ἐξαπίνης σημάντορος οὐ παρεόντος,     325
ὣς ἐφόβηθεν 'Αχαιοὶ ἀνάλκιδες· ἐν γὰρ 'Απόλλων
ἧκε φόβον, Τρωσὶν δὲ καὶ "Εκτορι κῦδος ὄπαζεν.

A pair of similes at 15.263ff. marks Apollo's restoration of Hektor's status from wounded warrior to vigorous fighter. The first, the famous repeated horse simile, conveys the hero's physical reanimation and sudden sense of freedom. The second simile (271ff.) sets the internal scene in the forest: men and dogs chase their quarry. The first concentrates on the individual warrior, while the second unfolds the wider context by summarizing the massive counterattack of Danaans, Hektor's opponents (ὁμιλαδόν, 277). The second simile begins in an ambiguous manner by comparing the Danaans to hounds and hunters pursuing a stag or wild goat. The comparison raises the false expectation that Hektor, like the hunted quarry, will fall victim to his pursuers, as is indeed the case in the simile. But a sudden shift in emphasis in the second simile draws our attention back to Hektor. A well–bearded lion appears, stealthily, amidst the shouting and barking. The dogs and men flee from their quarry: ... οὐδ' ἄρα τέ σφι κιχήμεναι αἴσιμον ἧεν (274). It was not ordained for them to apprehend their prey. Before, the men were eager; but now their courage fails them. The resumptive clause comprises two verses which describe the throngs attacking with swords and spears, corresponding to the active pursuit of the hunters before the lion appears; the following two verses expose their fear upon seeing Hektor and mirror the startling intrusion of the lion. The simile successfully has the audience experience the Danaans' surprise at Hektor's return by reversing the expectation of the hunting simile, which shifts to the common scene of a lion scaring off the men and dogs well known from herding similes. The simile is thus a combination of the hunting and marauding lion types.

Thoas, the best of the Aitolians, remarks on Hektor's unexpected return to the battlefield and surmises that some god has rescued Hektor: "for not without Zeus whose thunder roars does this chieftain rage thus" (292f.). Thoas then appeals to his comrades to resist his onslaught, and they rally to the cause. Sixteen lines of narrative, plus another lion simile (323ff.) mark the successful routing of the Achaians.[19] Mortal and god are compared to two lions (θῆρε δύω, 324) who stampede flocks of cows

---

[19] The rout is emphasized by ἐφόβηθεν in the resumptive clause, cf. ἧκε φόβον (327). A catalogue of slain warriors follows at 328ff.

and sheep. They invade suddenly in the dead of the night when no herdsman is about: ἐλθόντ' ἐξαπίνης σημάντορος οὐ παρεόντος (325). The simile emphasizes what happens when Hektor attacks in tandem with Apollo: the Achaians lose heart. The passage confirms Thoas' suspicion that Hektor operates in complicity with some god, as Apollo enshrouds Hektor with a fleecy cloud and himself shakes the aegis. The magical shield acts a kind of mask with animal attributes. Its epithet (ἀμφιδάσειαν, 309), recalls the mane (ἠυγένειος) of the marauding lion in the simile at 275. Like Agamemnon's telamon, the aegis has the same effect of bringing fear (φόβον, 310) to men as the lion does in the simile: ὣς ἐφόβηθεν Ἀχαιοὶ ἀνάλκιδες (326).

From the moment of Hektor's conspicuous reappearance marked by the first simile of the lion as hunter (271ff.) until the catalogue of slain warriors (328ff.) appended to the second lion simile, the atmosphere of the lion image is sustained. The image of the cowardly victims (stags) of the first simile is resolved in the second simile by the emotive phrases θούριδος ἀλκῆς (322) and ἀνάλκιδες (326) embracing the simile. Both describe the deer as a symbol of cowardice. The fear of wild and domestic animals before the lion, like the fear of the Achaians before Hektor, allows the aggressor free reign to triumph. The word ἀνάλκιδες has a double meaning. Literally, it means "without defense," as stated in the simile. No shepherd is about to defend them. Figuratively, it spells the Achaians' cowardice.[20]

In the second half of Book 15 the imagery conspires to keep alive the overriding image of Hektor as a wild animal. Lion imagery recurs in Book 15 in three similes, two developed (586ff., 630ff.), and one short (592). The lion identification shifts from Apollo/Hektor, to an Achaian warrior (Antilochos), to the Trojans at large in the short simile Τρῶες δὲ λείουσιν ἐοικότες ὠμοφάγοισι (592), and back again to Hektor and a god (Zeus) at 630ff. Dramatically the short simile marks a change in the tide or war: the Trojans gain the upper hand in the fight beside the ships. Zeus deceives the Argives and grants glory to Hektor (596).

A chain of similes follows to chart the Achaians' gradual loss of courage, as a glance at the resumptive clause of each reveals. First, in a

---

[20] At 4.234 Agamemnon rebukes the Argives by telling them not to forget their furious valor, and later at 242f. he upbraids them, comparing them to fawns in whom there is no valor (245). Similarly Poseidon–Mentes likens the Trojans to fawns who are ἀνάλκιδες (13.104).

passage marked by a wind simile the Danaans remain steadfast and do not flee (618ff.); another wind simile (624ff.) follows as they begin to lose heart, and their spirits waver (629); and finally the Achaians flee in terror like myriad cows before a lion (636ff.). This final reversal occurs in the resumptive clause of a marauding lion simile, which as elsewhere, emphasizes rout.

The internal defender, the shepherd, is called "inexperienced" (632). In the earlier lion simile involving Hektor at 15.323ff. the herdsman, as we saw, is said to be absent: σημάντορος οὐ παρεόντος (325). This explicit reference to the defender is unusual and may be an oblique reference to Achilles' absence, noted at 399ff. (cf. 64ff.). Both similes furthermore mark an attack by Hektor in collusion with a god. Before, Hektor attacked with Apollo; now the Achaians flee before Hektor and Zeus. The two similes sequentially measure Hektor's progress in the offensive against the ships and comment on his near success in the absence of Achilles.

Hektor's majesty, indeed his quasi–divine status, is evident in the narrative preceding the simile chain. A short alternative simile compares him to Ares or a fire raging in the depths of the thick forest (605ff.). In the imagery of the succeeding lines, there are allusions to an animal.

> μαίνετο δ' ὡς ὅτ' Ἄρης ἐγχέσπαλος ἢ ὀλοὸν πῦρ 15.605
> οὔρεσι μαίνηται βαθέης ἐν τάρφεσιν ὕλης·
> ἀφλοισμὸς δὲ περὶ στόμα γίγνετο, τὼ δέ οἱ ὄσσε
> λαμπέσθην βλοσυρῆισιν ὑπ' ὀφρύσιν, ἀμφὶ δὲ πήληξ
> σμερδαλέον κροτάφοισι τινάσσετο μαρναμένοιο
> Ἕκτορος· αὐτὸς γάρ οἱ ἀπ' αἰθέρος ἦεν ἀμύντωρ 610
> Ζεύς, ὅς μιν πλεόνεσσι μετ' ἀνδράσι μοῦνον ἐόντα
> τίμα καὶ κύδαινε...

A slaver (ἀφλοισμός) drools from the lip; the eyes beneath the brow gleam; σμερδαλέος describes the helmet shaking thunderously about the temples of the warrior's head. Hektor is more a berserk than merely an angry man. Three details in the description evoke a wild animal, lion or boar. ἀφλοισμός, a *hapax legomenon,* recalls the foam oozing from the mouth of the lion in our model simile (20.168). The burning eyes of a lion are described in an *Odyssey* lion simile;[21] the detail ὑπ' ὀφρύσιν "under the brow" perhaps suggests the prominent brow and sunken eye of the lion. This feature of the lion's physiognomy is singled out in the simile at 17.

---

[21] *Od.* 6.130f.; cf. 20.172, γλαυκιόων.

133ff.[22] The suggestion of a fiery glare in 608 is reinforced by the earlier simile mentioning fire, which comes from the "depth of the thick forest," the haunts of wild beasts. Finally, Hektor is said to rage, μαίνεται, cf. μαίνηται (605f.).

One should not be surprised to find residual lion imagery in the present passage. The short lion simile at 592 comparing the Trojans at large with lions establishes the identification between lions and the Trojans, whose leader, Hektor, has been compared to a lion earlier in the Book. The narrative situation, as we saw in Book 12, invites a marauding lion simile: the aggressor, eager to penetrate a barricade verges on routing his opponents. The language in 615–618 moreover contains two phrases from Hektor's lion–boar simile in the *Teichomachia*: στίχας ἀνδρῶν πειρητίζων (cf. 12.47), and the curious use of πυργηδόν for the serried ranks of men (cf. 12.43). Indeed a marauding lion simile does occur in the last of the chain similes, and it marks the rout.


## Conclusion

The foregoing examples demonstrate that in composing lion similes the poet worked from a standard but flexible model, framed by an introduction and conclusion, with the internal development relying on an alternation of verses or phrases that portray the action and reaction (or character) of the attacker and victim. The introduction begins with the word preceding the simile and is typically an action verb or emotive term. This initial element is followed by a conjunction and the word λέων or synonyms, which announce the simile subject. Occasionally the victim is announced first, e.g., 11.172ff. The internal development begins by stating the respective intent or motive of the lion or his adversaries to attack and continues with the actual encounter. The motives of the lion and farmers are represented from the points of view of each, and the offensive and defense are treated as separate elements. Less frequently the motive may be omitted from the internal development but included in the resumptive clause to explain the warrior's corresponding motive (e.g., 5.143). The elements of motive, attack, and defense do not always appear in the same sequence; the composer can shift, modify, or amplify them according to what has preceded or what will follow in the narrative. For example, after the wounding

---

[22] Cf. p. 44.

of Diomedes in his *aristeia*, the first element after the introduction is the shepherd's strike, resulting in the lion's wounding (5.136ff.). The poet most frequently develops his simile by amplifying the ethos of the animal attacker (the wounded lion grows angry) or the pathos of the victim. The conclusion of the simile summarizes or echoes either the first or final element of the internal development. The type is so well established that the image of the lion may be sustained in the immediate vicinity of explicit lion similes.

# Chapter V
# Pursuit in Battle: Hunting Similes

## Characteristics and Affinities with Lion Similes

There are eighteen developed hunting similes in the *Iliad*.[1] After marauding lion similes, hunting similes are the second most extensive type in the war epic. In addition to their apparent popularity, hunting similes resemble marauding lion similes in internal pictorial content and their poetic functions in relation to the narrative. The naming of lion and boar as alternate subjects in similes indicates that the epic tradition regarded the boar, the most frequent object of the hunt, as somewhat interchangeable with the king of beasts,[2] with the important difference, however, that the boar never kills an opponent. The adversaries of the hunted quarry are the hunters and his hounds. These are, of course, equivalent to the defenders in the lion similes, the shepherd and his dogs. The formulaic phrase κύνες θαλεροί τ' αἰζηοί is equally at home in a lion or a boar simile (3.26, cf. 11.414). The practice of attributing anthropomorphic states to the lion extends to the boar who shares the same emotions and internal states with the lion. The lion and boar are ὀλοόφρων (15.630, cf. 17.21), both trust in their strength, ἀλκὶ πεποιθώς (5.299, 17.61, cf. 13.471, 17.728); the boar, lion, and leopard neither fear nor feel dread ([οὐ] ταρβεῖ οὐδὲ φοβεῖται), and so forth.[3] Despite the sharing of emotive vocabulary between marauding lion and hunting similes, there is a greater emphasis in hunting similes on action and on physical qualities than on emotive states. Whereas lion similes tend to repeat the emotive state of the lion in the resumptive

---

[1] 3.23ff. (lion vs. deer or goat), 8.337ff. (lion or boar), 10.360ff. (deer or hare), 11.292ff. (lion or boar), 11.324ff. (boar), 11.414ff. (boar), 11.474ff. (deer), 12.146ff. (boar), 13. 198ff. (lion vs. goat), 13.471ff. (boar), 15.271ff. (deer or goat), 15.579ff. (fawn), 17.281ff. (boar), 17.725ff. (boar), 21.573ff. (leopard), 22.189ff. (deer). Two similes mention a lioness protecting her cubs from the hunter (17.133ff. and 18.318ff.).

[2] Lion or boar: 5.299, 782, 7.256, 11.292ff., 12.41ff.; 17.20ff. includes lion, boar, and leopard.

[3] 12.46 (lion or boar), 21.575 (leopard), cf. σθένεϊ βλεμεαίνων (-ει) 17.22, 135 (lion), 12.42 (lion or boar).

clause, hunting similes show a fondness for repeating action verbs from the simile in the resumptive clause; certain verbs, such as σεύω, are used in more than one simile.[4]

Hunting similes principally mark pursuit and flight, and steadfast defense, with action verbs corresponding with the ideas of pursuit, attack, and flight.[5] It is much less common, however, for a hunting simile to be triggered by an introductory verb of action, as often happens in lion similes. The hunting simile simply begins by announcing the hunter and hounds and then names the object of the hunt. Sometimes the object of the hunt is announced first, and an action verb follows. Then the circumstances of the chase are given—usually, but not always parallel to the narrative context. If a warrior is wounded, the wounding of the object of the hunt may be described (e.g., 11.474ff.); if the pursued warrior stands up to his attackers, we usually find the boar wheeling on the dogs and men (e.g., 17.725ff.), or the opposite situation of the hunters standing up to the boar (11.414ff.). Although hunting similes tend to begin from the point of view of the hunter and his hounds, the perspective often changes to the object of the hunt by describing the reaction of the pursued animal or the outcome of the chase. In the absence of explicit emotive words, the shift of attention to the victim is what creates the pathos of the simile. Hunting similes are characterized by reversals and unexpected events. For example, when the Trojans crowd around the wounded Odysseus in Book 11 a simile begins by announcing the jackal subject. It names a deer as the object of a hunt that is to be imagined in progress. The simile then changes from the perspective of the jackals to the deer; we hear the circumstances of its wounding by a hunter's arrow, its attempted flight, and finally its death as the jackals make ready to attack the carcass. Suddenly attention is drawn to a ravening lion sent by a spirit—the simile's equivalent of a divine intervention.[6] As with the hunter, the identity of the δαίμων is left mysterious. The jackals flee while the lion scavenges on the corpse. The simile, with its two repeated sets of action verbs can be analyzed as follows. Note how the internal scene extends beyond the resumptive clause (482–84).

---

[4] σεύω occurs in the simile and resumptive clause twice (11.292ff., 11.414ff.); ἀίσσω: 11.417, 17.726, 734; θρῴσκω: 15.580, 82; διατρέω: 11.481, 486.

[5] Cf. Krischer's *Flucht und Verfolgung, Standhalten, Zurückweichen.*

[6] G. E. Markoe, "The 'Lion Attack' in Archaic Greek Art: Heroic Triumph," *CA* 8 (1989) 89, cites this simile in support of his theory that the lion attack in Homer is divinely inspired.

"Ως εἰπὼν ὃ μὲν ἦρχ', ὃ δ' ἅμ' ἕσπετο ἰσόθεος φώς.   11.472
εὗρον ἔπειτ' Ὀδυσῆα Διὶ φίλον· ἀμφὶ δ' ἄρ' αὐτὸν
Τρῶες ἕπονθ' ὡς εἴ τε δαφοινοὶ θῶες ὄρεσφιν
ἀμφ' ἔλαφον κεραὸν βεβλημένον, ὅν τ' ἔβαλ' ἀνὴρ        475
ἰῶι ἀπὸ νευρῆς· τὸν μέν τ' ἤλυξε πόδεσσι
φεύγων, ὄφρ' αἷμα λιαρὸν καὶ γούνατ' ὀρώρηι·
αὐτὰρ ἐπεὶ δὴ τόν γε δαμάσσεται ὠκὺς ὀιστός,
ὠμοφάγοι μιν θῶες ἐν οὔρεσι δαρδάπτουσιν
ἐν νέμει σκιερῶι· ἐπί τε λῖν ἤγαγε δαίμων        480
σίντην· θῶες μέν τε διέτρεσαν, αὐτὰρ ὃ δάπτει·
ὡς ῥα τότ' ἀμφ' Ὀδυσῆα δαίφρονα ποικιλομήτην
Τρῶες ἔπον πολλοί τε καὶ ἄλκιμοι, αὐτὰρ ὅ γ' ἥρως
ἀίσσων ὧι ἔγχει ἀμύνετο νηλεὲς ἦμαρ.
Αἴας δ' ἐγγύθεν ἦλθε φέρων σάκος ἠύτε πύργον,        485
στῆ δὲ παρέξ· Τρῶες δὲ διέτρεσαν ἄλλυδις ἄλλος.
ἤτοι τὸν Μενέλαος ἀρήιος ἔξαγ' ὁμίλου
χειρὸς ἔχων, εἷος θεράπων σχεδὸν ἤλασεν ἵππους.

The adversative αὐτάρ in the resumptive clause of the simile (483), followed by διέτρεσαν (486), corresponds to αὐτάρ in the simile (478), which is followed again by διέτρεσαν (481). The parallelism reflects the fact that the simile relates back to the preceding narrative situation of the Trojans crowding around the corpse, just as it looks forward to the extraordinary and unexpected self–defense of Odysseus. There are in essence two separate similes joined by ring composition. The first, concerning the jackals and deer, actually misleads us by describing the pathetic terms of the wounding of the deer now crowded around by scavengers. We are led to suppose that Odysseus too will fall to the Trojans. But our expectation is thwarted by the second part of the simile, which is really a variation on the marauding lion type. The lion approaches, scatters the wild dogs as he does the sheep dogs, and feasts on the carcass. In defending Odysseus Aias scatters the Trojans (486) so that the attention shifts from the threatened hero to the fleeing Trojans. The hunting dog correspondingly becomes the lion's prey (481). In this and other hunting similes the poet exploits the uncertain outcome that attends every hunt, and he draws us into the suspense by shifting the perspective from hunter to hunted victim. This variable point of view, which we have seen in lion similes, may occur within a simile, in sequences of similes as in Book 11, or in a sustained network of imagery such as that in the climax of the epic when Hektor is transformed from hunter to prey.

Hunting similes also differ from lion similes in the emphasis placed on the dog. Many hunting similes focus on the hunter's hounds, while the hunter is suggested only by metonymy. The hound in the hunting similes is the active counterpart of the undifferentiated sheep dogs of the marauding lion similes and the metaphorical dogs in scavenging threats. At one extreme the dog in Homer is a utilitarian creature; at the other, a loathsome scavenger. The polarity carries through in both epics. But whereas the dog appears frequently in the action of the *Odyssey* narrative, in the *Iliad* he is virtually absent, surfacing only in the similes.

### The Roles of the Hound and the Boar

A brief contrast between the dog and horse in the *Iliad* is instructive for understanding the poetic functions of the canine.[7] So different are the uses of these two animals that they have been called the "Caliban" and "Ariel" of the Homeric animal world.[8] Whereas the horse is a ubiquitous and indispensable actor in the plot, the dog remains invisible throughout the *Iliad*. Homeric horses are flesh and blood creatures with a will and a way of their own, individuals honored by a rich system of epithets and sometimes accorded names and a lineage.[9] The dog is indistinct, a member of a pack, with few epithets, summoned only in the similes, or more frequently in metaphorical contexts. Dogs appear in narrative action only twice, once in the plague (1.50) and once on Patroklos' pyre (23.173).[10] From a compositional point of view, then, the horse in the *Iliad* is largely

---

[7] See C. Mainoldi, *L' image du loup et du chien dans la Grèce ancienne d' Homère à Platon* (1984) 97–126; S. Lilja, *Dogs in Ancient Greek Poetry* (1976).

[8] Redfield, *Nature* 195.

[9] Delebecque, *Cheval* 137–161, and *passim*.

[10] For an interpretation of why Apollo shoots the mules and dogs first at 1.50, see M. Mühl, *REG* 71 (1971) 1–16. There is a higher occurrence of scavenging threats in the *Iliad* than in the *Odyssey*, simply because the subject–matter of the *Iliad* will lend itself to the inclusion of such threats. Yet five such threats do occur in the *Odyssey* (*Od.* 14.133, 18.87, 21.363, 22.476, cf. *Od.* 24.292), and as in the *Iliad* they appear with greater concentration at the climax of the epic. But the only time that the threat is actually carried out is in the *Odyssey*, when Telemachos, Philoitios, and Eumaios hack away at Melanthios' genitals to feed them to the dogs (*Od.* 22.476).

restricted to the narrative.[11] The dog figures in a flexible network of imagery comprising the hunting similes, similes of shepherds and sheep dogs, the theme of the scavenger feeding on the abandoned corpse, and the various words of abuse derived from κύων. As a term of abuse, the vocative κύον is the gravest insult, used with some consistency by Achilles to vent his anger against Agamemnon.[12] The metaphorical association between the bitch in heat, signified by κυνῶπις (*Od.* 4.145, 8. 319) and κύον ἀδέες (*Od.* 19.91), and shameless female behavior in the *Odyssey* is paralleled in the *Iliad* by a connection between Helen and the house of Priam. By virtue of her association with Paris Helen has laid the palace of Priam open to moral censure, and the Trojans by extension are regarded as having fallen prone to an excessive appetite of a different kind, battle lust.[13] The tones of association reflect the varying degrees of domesticity inherent in the types of dog: there are marked variations between the table dog and the hunting hound, or the sheep dog and the scavenger, or indeed between the watch–dog and the rabid dog. The distinctions emerge sometimes only ambiguously, especially when juxtaposition of different types occurs. For example, in a simile sequence from Book 11 discussed below, Hektor is compared first to the hunter, next to the audacious hound, and in the ensuing passage is slanderously called "a dog."

The importance of the dog's function lies in mediating between the realm of the hunt, the natural world, and the battlefield. There are three possible spheres of pursuit. In the peace–time realm of the hunting similes the hunter, along with his hounds, pursues the boar primarily for sport. In the animal kingdom the predator attacks his prey, while the scavenger (including the jackal or wolf, the dog's forebears) feeds on the carrion. In the arena of war man hunts man and the scavenger feeds on the abandoned corpse. The dog stands on man's side in the artificial world of the hunting simile; in war the dog stands apart from man, ready to scavenge. The realm of nature and the arena of war are not far removed from one another: man potentially falls prey to his own kind. Hektor and Achilles are identified with the dog who serves to show the bloodthirsty ebb and flow in the

---

[11] Horses are simile subjects at 6.506–511=15.263–68, 22.22f., 162ff. They are the only creatures honored by similes: 2.764, 765 (cf. the brief horse catalogue at 2.761–70, prefaced by an invocation to the Muse), 16.384ff.

[12] M. Faust, *Glotta* 48 (1970) 28.

[13] The metaphorical uses of the dog carry moral allusions to surfeit. In ten out of thirty scavenging threats κόρος, "surfeit," (or a related form) occurs in the threat or in the immediate context, e.g., κορέεις κύνας ἠδ' οἰωνοὺς | δημῷ καὶ σάρκεσσι (13.831f.).

hearts of adversaries of the poem. The pejorative connotations of the dog's character that build up during the course of the poem culminate in Book 22, where canine references abound.

The majority of the hunting similes report the clamorous climax of the hunt when the huntsmen and their hounds confront the wild boar. Despite his ungainly shape the hulking 180–pound boar is an agile and swift creature with a keen sense of hearing and smell. When cornered he stands his ground before the baying dogs. In a flash the creature may rush forward, scattering the dogs, and head straight for the human attackers. As the Townleian scholiast observes on 17.725–29, στρέφεται γὰρ τοῦτο μόνον τῶν ζῴων διωκόμενον, ὁ κάπρος. A simile from Book 12 vividly describes the uproar caused by the tusks of the boar as he bulldozes the thicket and underbrush in his onslaught (12.146ff.). The boar in a sense is a warrior. He has the armor: a hide as tough as a carapace and surmounted by a bristling crest, and two tusks for spears. And in conflict he behaves like a fighter. He stands his ground and makes an attack in spite of overwhelming numbers of men and dogs. Because of the boar's own courageous nature and the daring required of the hunter to confront this dangerous beast, the boar–hunt metaphorically expresses the heroism with which the warrior defends himself. The beast more often than not wheels upon his attackers and turns them to flight. But some boar similes focus on the dogs alone, as we participate, for example, in the hound's desire to rip off a small piece of hide from the flank of the boar (8.338ff.).

## Hunter and Hounds: Hektor and the Trojans: Books 11 and 17

Except for two similes comparing individual Achaians to hounds, all hunting similes have as their subject of comparison Hektor, or Hektor and the Trojans.[14] These are closely allied to the similes of farmers warding off the attack of lions. There, too, the Trojans are compared to the men and dogs.[15] The general tendency to compare the Trojans to huntsmen and shepherds and their dogs corresponds to a propensity to equate the Achaians with the boar and the lion, the two beasts who respectively give sport to the hunter and harass the shepherd. The marks of civilization often im-

---

[14] Antilochos: 15.579ff.; Achilles: 22.189ff.

[15] Again, there is an exception: the Aiantes are compared to shepherds attempting to avert the attack of a lion (18.161f.).

plied by hunting for sport waver as Hektor's status fluctuates between that of the hunter and hound. The civilized connotations vanish when Hektor becomes like an animal pursued by Achilles.

A pronounced comparison between Hektor and his troops and the hunter and hounds occurs in Book 11 beginning with the simile at 291ff.:[16]

> Ὣς εἰπὼν ὄτρυνε μένος καὶ θυμὸν ἑκάστου.
> ὡς δ᾽ ὅτε πού τις θηρητὴρ κύνας ἀργιόδοντας
> σεύηι ἐπ᾽ ἀγροτέρωι συὶ καπρίωι ἠὲ λέοντι,
> ὣς ἐπ᾽ Ἀχαιοῖσιν σεῦε Τρῶας μεγαθύμους
> Ἕκτωρ Πριαμίδης βροτολοιγῶι ἶσος Ἄρηι.

This simile is the initial member of a sequence of four similes in Book 11 describing the hunt. Together the four similes mark the Trojans' successful efforts to turn the tide of war after Agamemnon's wounding. In the *aristeia* of Agamemnon, as we saw, animal imagery is almost wholly restricted to the lion; but now with the advent of Trojan successes, hunting imagery takes over, and Hektor as leader of the troops is appropriately compared to the hunter in the first simile. This neat shift in imagery, corresponding as it does to a narrative reversal, contributes to the tight construction of the battle poetry in Book 11.[17]

After the wounding of Agamemnon, Hektor rouses the Trojan, Dardanian, and Lykian troops with a piercing battle–cry. He beseeches them to pluck up their courage and breathes strength into the breast of each man. The simile compares him to a hunter, τις θηρητήρ, driving his white-toothed hounds against a lion or boar. "... So Hektor the son of Priam drove the great–hearted Trojans against the Achaians." The resumptive clause clearly equates the Trojans with the hounds. A short simile (295) comparing Hektor to Ares in the resumptive clause magnifies Hektor's stature.

---

[16] The association between Hektor and the hound and hunter has been evident as early as Book 8. Teukros complains to Agamemnon, "I cannot strike that rabid dog," κύνα λυσσητῆρα (299), referring to Hektor. He is then compared in a simile (338ff.) to a hound in pursuit of a boar or lion. This is an odd simile, functioning like similes of the lion stampeding a herd to mark the rout of an army, even though no group is mentioned in the simile. The boar and wasp similes (12.146ff.; 167ff. [sc. ἄνδρας θηρητῆρας, 170]) emphasizing the staunch defence of Leonteus and Polypoites at the gates provide other straightforward examples of the Trojans in general conceived as hunters and hounds. The association between Hektor and a dog in a simile recurs at 13.198ff.

[17] Cf. T. B. L. Webster, *From Mycenae to Homer* (1958) 232f., who analyzes the contrasts and cross–references in the similes in Book 11.

The Trojan attack is met by a counterattack. In the corresponding imagery a second hunting simile compares Diomedes and Odysseus, who have roused one another to fight off Hektor's attack, to boars turning the hounds to flight (324f.). The second simile is a simple reversal of the first. Now we catch a glimpse from the boar's point of view. The attacking men in the first simile are referred to as Τρῶας μεγαθύμους (294) and their leader as μέγα φρονέων (296); now the wheeling boars to whom the belligerent Achaian defenders are compared are called μέγα φρονέοντε (325).[18] The hounds in the second simile are referred to as κυσὶ θηρευτῇσι (325), a telescoping of κύνας and θηρητήρ from the previous simile; the men, who are conspicuously absent from the scene, are only suggested by metonymy. The Trojans are repulsed, but Hektor persists for a time, only to be forced to seek refuge in the throng of warriors when Diomedes wounds him. A dog insult follows. The proud Achaian shouts after him, "you have escaped death for now, you dog" (362). The term of abuse here has more than an isolated metaphoric innuendo: within the hunting sequence Hektor's status has fallen from that of huntsman to hound.

The hunt–motif continues to chart the progress in fighting. At 414ff. the Trojans, like a group of young huntsmen with their hounds (κύνες θαλεροί τ' αἰζηοί) surround Odysseus, now compared to a wounded boar. The point of view initially reverts to the hunters. Visually the simile helps to clarify the configuration of a mass of men surrounding a single warrior. The ring form is pronounced: the verb σεύω in the introductory and resumptive clauses emphasizes the overwhelming force of the attackers. But we also witness favorably the defense of the solitary victim. The scene homes in on the grinding action of the boar's white fangs in the crook of the jaw. An auditory detail is added, the gnashing sound of the teeth. The emphasis in the simile on the boar's tusks, his "offensive weaponry," corresponds to Odysseus' virtuoso spearsmanship. δουρικλυτός, one of Odysseus' epithets, occurs twice in the narrative preceding the simile at 396, 401 (cf. 11.661, 16.26). In the catalogue following the simile Odysseus wounds Deiopites in the shoulder with his sharp spear, stabs Chersidamas in the navel with a spear's stroke, spears Hippasos, and finally lunges his spear at Sokos. When the Trojans succeed in wounding

---

[18] Cf. 16.758, 824, where μέγα φρονέοντε occurs in a sequential pair of lion similes. μέγα φρονέων ἐβεβήκει occurs always after the C2 caesura (11.296, 13.156, 22.21); μέγα φρονέοντε stands in the same position, at the end followed either by πεσήτον (11.325) or μαχέσθον (16.758, 824), depending on the sense that is needed.

Odysseus they are compared to a pack of jackals feeding on a wounded stag (474ff.) in a simile analyzed above. There are no longer any men to dignify the scene, and the lonely forest setting where scavengers feed on carrion is a reminder of the vulnerability of the abandoned warrior on the battlefield. The degeneration from hounds to cowardly jackals implies that an inferior band of men has wounded a worthy warrior. We may observe in the overall network of lion and hunting imagery in Book 11 a regression of the Trojans from shepherds and hunters to hounds and watchdogs, and finally to scavengers.

The famous concluding lion–boar simile of Book 16[19] sets the mood for the fight over Patroklos' body in Book 17, which recalls the fight over the spoils from the Kalydonian boar. Simile and paradigm border on reality: the corpse in effect is treated like carrion. The metaphorical association between Trojans and hunters and hounds is intensified in Book 17. In all there are eleven references to the three main types of dogs, the scavenger, the watch–dog, and the hunting dog in this Book—a much higher frequency than in any other Book of the *Iliad*. The hunting and herding similes are closely interrelated and refer to the unifying action of the Book, the Achaians' concerted efforts to ward off the Trojans from the corpse.

Towards the beginning of the episode there are three lion similes which can be taken together. The text of the first two is given below:

Ἀτρείδης Μενέλαος ἐπεὶ κτάνε τεύχε' ἐσύλα.          17.60
Ὣς δ' ὅτε τίς τε λέων ὀρεσίτροφος ἀλκὶ πεποιθὼς
βοσκομένης ἀγέλης βοῦν ἁρπάσηι ἥ τις ἀρίστη·
τῆς δ' ἐξ αὐχέν' ἔαξε λαβὼν κρατεροῖσιν ὀδοῦσι
πρῶτον, ἔπειτα δέ θ' αἷμα καὶ ἔγκατα πάντα λαφύσσει
δηιῶν· ἀμφὶ δὲ τόν γε κύνες τ' ἄνδρές τε νομῆες          65
πολλὰ μάλ' ἰύζουσιν ἀπόπροθεν οὐδ' ἐθέλουσιν
ἀντίον ἐλθέμεναι· μάλα γὰρ χλωρὸν δέος αἱρεῖ·
ὣς τῶν οὔ τινι θυμὸς ἐνὶ στήθεσσιν ἐτόλμα
ἀντίον ἐλθέμεναι Μενελάου κυδαλίμοιο.

cf.

αὐτὰρ ὅ γ' ἐξοπίσω ἀνεχάζετο, λεῖπε δὲ νεκρὸν          17.108
ἐντροπαλιζόμενος ὥς τε λὶς ἠυγένειος,
ὅν ῥα κύνες τε καὶ ἄνδρες ἀπὸ σταθμοῖο δίωνται          110

---

[19] Schnapp–Gourbeillon, *Lions* 81–83, traces the interaction between the plot and the lion similes used of Patroklos in Book 16.

ἔγχεσι καὶ φωνῆι· τοῦ δ᾽ ἐν φρεσὶν ἄλκιμον ἦτορ
παχνοῦται, ἀέκων δέ τ᾽ ἔβη ἀπὸ μεσσαύλοιο·
ὡς ἀπὸ Πατρόκλοιο κίε ξανθὸς Μενέλαος.

The first lion simile compares Menelaos to a lion devouring a calf, a violent reversal of the opening simile at 4ff. where he is likened to a cow standing over her first–born calf. [20] The farmers attempt to scare him away with raucous yelling, but only from a distance. No one dares approach. Pale fear overcomes them: μάλα γὰρ χλωρὸν δέος αἱρεῖ (67). When Hektor leads an attack against the body, however, it is the courage of the lion in the next simile that turns to fear: τοῦ δ᾽ ἐν φρεσὶν ἄλκιμον ἦτορ | παχνοῦται ... (111f.), in stark contrast to the final line of the earlier lion simile (67). Now the spear–bearing shepherds and dogs boldly pursue as the lion backs away from the steading, slowly, step by step, disappointed of desire (109, 112).[21] The simile emphasizes the reluctance with which Menelaos must abandon the body; for Patroklos does in fact momentarily become an abandoned corpse, and Hektor and his men advance to strip off the armor. The Trojan leader lusts to "hack away the head" and give the body to the "Trojan bitches," (Τρῳῆσιν ... κυσί, 127), as if a hunter taking the prized parts of the animal for the trophy and opening the victim's entrails to reward the hounds.

Aias, realizing what has happened, rushes in to replace Menelaos; as he towers over the body with his shield he is likened in the third simile to a lioness protecting her whelps from the huntsmen in the forest (133ff.). The

---

[20] The meaning of the epithet κινυρή is uncertain. Perhaps it means emitting a plaintive or "threatening" (M. Leumann, *Homerische Wörter* (1950) 242, s. κινυρός) cry, because as a mother giving birth for the first time (οὐ πρὶν εἰδυῖα τόκοιο), she instinctively feels a need to protect her young.

[21] The word ἐντροπαλιζόμενος accurately describes the way that a lion retreats step by step, with his gaze trained on his opponents; he will continue to back off until he is a safe distance to make a getaway. The word has a special resonance in the *Iliad*. After speaking with Hektor Andromache turns away (ἐντροπαλιζομένη) with the greatest reluctance as she sheds a tear (6.496). Hera rebukes Artemis, and she tries to twist away (ἐντροπαλιζομένην) in hurt frustration as he boxes her on the ears, and her arrows fall clumsily about her (21.492). In the present simile (17.109ff.) Menelaos is compared to a frustrated lion shying away from the attack of men and dogs (ἐντροπαλιζόμενος); so he reluctantly leaves the corpse of Patroklos. Similarly, when Aias is checked in his rampage by Zeus, the frustrated warrior is compared in a short simile to a lion who retreats, ἐντροπαλιζόμενος (11.547). The uses of the words all have one point in common: a sense of helplessness in the face of enforced separation from what one desires most to be or be near: Andromache to Hektor; Artemis to be accepted by the male gods as a warrior; Aias to be the rampaging warrior; Menelaos to be the protector of Patroklos; and in the animal metaphor, the lion to possess his victim.

theme of the animal mother and her young intensifies the urgency of the situation and links the simile to the opening simile of the Book and the well–known simile at 18.318ff. comparing Achilles to a lioness sniffing out the tracks of the hunter who has lifted the cubs from the lair.[22]

Now that the corpse lies in its natural state, denuded and vulnerable, the threat of the scavenging dog becomes prominent. Along with the increasingly bloodthirsty tone of Book 17, the phrasing of the formulaic threat to the abandoned corpse alters perceptibly to make the ambiguous equation between Trojans and gluttonous dogs more suggestive. At 125–27 the gruesome reason for Hektor's removal of the armor is made known: Hektor wishes to lop off the head and feed the body to the "Trojan" bitches. Bitches, we know from Xenophon (*Cyn.* 3.1ff.), were especially prized as hunting hounds. The use of the feminine is thus accurate from the point of view of realism, while the metaphorical resonances of uncleanliness and promiscuity that the bitch evokes are ambiguously suggested. Τρῳῇσι(ν) κυσί(ν) recurs again at 255 (=18.179, Πατρόκλον Τρῳῇσι κυσὶν μέλπηθρα γενέσθαι) and at 17.272f. (δηίων κυσὶ κύρμα γενέσθαι Ι Τρῳῇσιν).

This last threat occurs in the short passage at 269–73 which marks Zeus' resolve not to allow the body of Patroklos to become prey to the dogs. He takes pity on his corpse, "since he had not hated the son of Menoitios, the companion of Achilles, when he was living" (270f.). The gentle, reasoning tone conveyed in these lines contrasts with the feverish struggle raging throughout the Book. To confuse the Trojans Zeus sends down a darkening mist upon their shining helmets.

Concurrently Aias succeeds in scattering the ranks of Trojans crowding around Patroklos' body. A simile (282ff.) compares Aias to a boar wheeling upon the hunters and dogs in a forest glen, as if the simile were an earthly confirmation of the divine decision. The identification between the boar wheeling on the "Trojan" dogs and Aias continues in the following battle narrative. Aias slays his victim with a spear which passes through the dog–skin helmet (κυνέης, 294).

The tempo of the battle over the corpse is maintained throughout the Book. Towards the end of the episode a magnificent hunting simile heralds the beginning of the long simile chain beginning at 725ff. This poetic *tour*

---

[22] Moulton, *Similes* 105–106, sees the simile in Book 18 as an intended reversal of the simile in Book 17.

*de force* celebrates the Aiantes' moment of victory as they lift the corpse out of the fray towards the ships.

> Ὣς ἔφαθ', οἳ δ' ἄρα νεκρὸν ἀπὸ χθονὸς ἀγκάζοντο
> ὕψι μάλα μεγάλως· ἐπὶ δ' ἴαχε λαὸς ὄπισθε          17.723
> Τρωικός, ὡς εἴδοντο νέκυν αἴροντας Ἀχαιούς.
> ἴθυσαν δὲ κύνεσσιν ἐοικότες, οἵ τ' ἐπὶ κάπρωι          725
> βλημένωι ἀίξωσι πρὸ κούρων θηρητήρων·
> ἕως μὲν γάρ τε θέουσι διαρραῖσαι μεμαῶτες,
> ἀλλ' ὅτε δή ῥ' ἐν τοῖσιν ἐλίξεται ἀλκὶ πεποιθώς,
> ἄψ τ' ἀνεχώρησαν διά τ' ἔτρεσαν ἄλλυδις ἄλλος.
> ὣς Τρῶες εἷος μὲν ὁμιλαδὸν αἰὲν ἕποντο          730
> νύσσοντες ξίφεσίν τε καὶ ἔγχεσιν ἀμφιγύοισιν·
> ἀλλ' ὅτε δή ῥ' Αἴαντε μεταστρεφθέντε κατ' αὐτοὺς
> σταίησαν, τῶν δὲ τράπετο χρώς, οὐδέ τις ἔτλη
> πρόσσω ἀίξας περὶ νεκροῦ δηριάασθαι.

The sense of triumph accompanying this feat emerges in the details of the simile. Our sympathy lies entirely with the boar: the ferocious beast lies wounded. We can imagine the confidence of the hounds as they rush, eager to rip apart the flesh: ἕως μὲν γάρ τε θέουσι διαρραῖσαι μεμαῶτες (727). But the boar recovers, trusting in his mighty strength, ἀλκὶ πεποιθώς; he turns the men and dogs away in confusion. At a time when Zeus had been favoring the Trojans[23] the Aiantes accomplish their task against expectation with sheer brute force and physical superiority. For the moment the Trojans are powerless hounds.

In the boar simile the comparison lies between the Trojans and simply "the dogs" (κυνέσσιν, 17.725); as in other examples, the hunters recede into the background. The specific comparison between Trojans and dogs is apt here. The narrative verse before the simile focuses on the clamor of the throng: ἐπὶ δ' ἴαχε λαὸς ὄπισθε | Τρωϊκός (17.723f.). The soldiers shriek as they see the corpse lifted up by the towering pair of warriors; deprived of the object that has so keenly whetted their appetite, they are almost like table dogs straining for a scrap of food out of reach in their master's hand. The final hunting simile augments the earlier simile comparing Aias with a boar wheeling upon the men and dogs (281ff.) and also reca-

---

[23] Cf. Menelaos' remark at 17.688f., πῆμα θεὸς Δαναοῖσι κυλίνδει, | νίκη δὲ Τρώων.

pitulates the prominent identification in the Book between Trojans and dogs, both hunting hounds and scavengers.

In Book 17 the opposing sides are pitted against one another for the possession of the corpse. The attractiveness of the body as booty is enhanced by the superb armor of Achilles still enshrouding it. At 17.122 we learn that Hektor has stripped the armor from the corpse. The helmet and shield are comparable to the prized parts of the boar, the head and the hide. Once they are removed the body itself becomes a coveted possession, treated almost as an animal trophy in accordance with Hektor's wish to hew the head from the body. In Book 18, when the scene flashes from Achilles' lament back to the struggle over the corpse, Hektor comes close to fulfilling his desire. Iris plummets to the battlefield to alert Achilles. We are impressed by how savage the civilized hunter has become, as Iris reports Hektor's obsession with possessing the corpse that he may display his kill: "The anger within his heart is urgent to cut the head from the soft neck and set it on sharp stakes"—followed by the more pointed phrasing of the dog threat, Τρῳῇσιν κύσιν (18.179). Later Achilles will exhibit similar behavior towards Hektor.

# Chapter VI
# Pursuit and Attack: Reversals in Hunting Imagery

A noticeable reversal in the animal imagery occurs once Achilles appears on the scene. Hektor is no longer dignified by comparisons to the hunter, nor to any attacking force, except in the eagle–hare simile in the chase scene. The body of Patroklos, who before was Achilles' hunted counterpart, is rescued from the fate of becoming a Trojan trophy. Hektor now loses his status; he becomes the hunted victim of Achilles. In the course of gaining revenge Achilles is repeatedly likened to pursuers and attackers: the hound (and allusively the hunter), wild dogs (wolf, scavenging or rabid dogs), and finally, the lion. Fire and star imagery figures prominently throughout.[1] The poet draws for analogy on all possible spheres pertaining to destructive forces in order to intimate the intensity and scope of Achilles' revenge. The hero's appropriation of Hektor's metaphorical guise, answering to Hektor's literal borrowing of Achilles' armor, implies the all–consuming hunger underlying his revenge.

The tendency to compare Hektor now to the hunter, now to the hunting hound is transferred to Achilles, who is associated with dogs: the rabid dog, Orion's dog Sirios, an anonymous hunting dog, and the dreaded scavenger. In the duel scene Achilles expresses his desire to eat Hektor's flesh raw like a scavenger; he loses his heroic stature and becomes something less than a man. As for Hektor, his state is utterly debased; he becomes the meat to glut the voracious appetite of the scavenger.

The various identifications between Achilles and attacking forces drawn from animal imagery are set forth below. In considering the progression of imagery in the Books leading up to the duel between Hektor and Achilles it will be convenient to deal first with imagery of pursuit, and second with imagery of attack. There are in essence two attacks in 22: the duel itself, and Priam's empathetic vision of the dogs defiling his body.

---

[1] See C. H. Whitman, *Homer and the Heroic Tradition* (1958) Ch. 7, "Fire and Other Elements."

Imagery of pursuit:

1.  18.318ff.: Achilles is compared in a simile to a lioness whose young have been stolen by a huntsman, ἐλαφηβό-λος.

2.  18.334ff.: Achilles tells Patroklos he will not bury him until he brings the head and armor of Hektor and beheads the twelve Trojans.

3.  20.164ff.: Achilles is compared in a simile to a lion threatening the welfare of an entire community.

4.  21.27ff.: Achilles seizes twelve Trojan youths, compared to fawns.

5.  21.252: Achilles is compared in a simile to an eagle whose epithet is "the hunter" (τοῦ θηρητῆρος).

6.  21.573ff.: Achilles is compared in a simile to a huntsman facing a leopard.

7.  22.189ff.: Achilles is compared in a simile to a dog chasing a fawn.

Imagery of attack:

8.  21.542f.: Achilles' breast is seized by λύσσα.

9.  22.26ff.: Achilles is compared in a simile to Orion's dog. (Followed by Priam's references to scavengers: 42, 66, 75, 89.)

10. 22.339, 345, 348, 354: Scavenging threats and terms of abuse in the duel scene. (Implied: 22.347, 371ff.; other scavenging threats: 22.509, 23.183ff.)

11. 22.262ff.: Achilles' parable to Hektor. "As there are no oaths between man and lion..."


## Imagery of Pursuit: Books 18–21

At the moment Achilles makes it known that he will re–enter the fighting a reversal in the animal imagery occurs that involves two lion similes (18.161f.; 318ff.) and allusions to Hektor and Achilles as hunters. Iris warns Achilles of Hektor's obsessive desire to haul away the body of Patroklos. She arouses his anger, warning him of Hektor's wish to exhibit his kill by setting the head on a stake (18.176f., cf. 17.126f.). The fight over the boar trophy in the Meleager paradigm is thematically transferred to the fight over the corpse of Patroklos, who was compared at his death to a slain boar. A few lines before Iris' speech a lion simile has confirmed and reinforced Hektor's intentions. The Trojan leader is compared to a hungry lion feeding on a carcass (18.161ff.). "As herdsmen ... are not able to frighten away the tawny lion ... so the two Aiantes ... were not able to scare Hektor ... away from the body." The timely intervention of Iris, however, saves Patroklos' corpse from an ignominious fate.

Achilles now expresses a longing similar to Hektor's towards Patroklos' corpse. He promises his shade that he will refrain from burying him until he brings "to this place the armor and head of Hektor" (18.334). He promises further to behead twelve Trojan youths at his funeral pyre. Like Hektor in his wish to seize the armor and display the head of Patroklos, Achilles now believes that in erecting a trophy of the prized parts of the murderer, he will gratify his friend. Again the parts of the enemy in question are the head and the armor, corresponding to the head and hide of the animal trophy. The capture and sacrifice of the twelve Trojan youths resemble the hunting and sacrifice of animals. The youths, who are compared in a short simile (21.29) to fawns, are to be immolated along with sheep, cattle, dogs, and horses (23.166–174). Achilles' promise to sacrifice the twelve Trojans and to display the head and armor of Hektor is a reprisal of Hektor's desire to erect a trophy.

A lion simile at 18.318ff. mentioning a hunter reinforces the reversal. Like Hektor's lion simile it takes place a few lines before the speech telling of the wish to possess and display the body. The simile insists upon the theme of illegal ownership: the theft of the lioness's young by a deer hunter. The lioness's revenge is chillingly implied. Bitter anger seizes the mother: μαλὰ γὰρ δριμὺς χόλος αἱρεῖ, and scouring the glens she sniffs out the tracks of the huntsman, εἴ ποθεν ἐξεύροι. The simile ends on an inconclusive and suspenseful note. Occurring in the context of ritual mourning the simile, like the short smoke simile (23.100) marking Achilles' attempt to embrace Patroklos' fugitive shade, comments upon the anxiety of the mourner to retrieve the cause or object of his loss in order to make reparation. Although the introductory and resumptive clauses of the simile repeat the outward manifestation of Achilles' mourning (πυκνὰ μάλα στενάχων ... ὥς ὁ βαρὺ στενάχων), the simile concentrates on the internal trauma and vengeful pursuit of the bereft lioness.

The simile also has an important bearing on the plot. It appears at the moment Achilles renounces his hurt pride to avenge the death of his comrade. The simile is thus pivotal to the action of the poem. An earlier simile (17.132) mentioned above compares Aias, as he protects the body of Patroklos from Hektor and his "Trojan hounds" (17.125), to a lioness shielding her young from huntsmen. Now a deer hunter has stolen the young of the lioness. The identification between Hektor and the hunter is pronounced through the specific use of the adjective ἐλαφηβόλος in the singular modifying ἀνήρ (18.319); it is further enhanced by the recent allusion in Iris' speech to Hektor as a hunter. The wrath of Achilles is now mobilized: he is ready to avenge himself, to capture Patroklos' murderer, as the lioness will sniff out the hunter's tracks. In the concluding line of the simile one may even note the inclusion of vocabulary appropriate to the wrath theme: μάλα γὰρ δριμὺς χόλος αἱρεῖ (18.322).[2]

The comparison between Achilles and a wild beast continues in Achilles' monumental lion simile at 20.164ff., which amplifies the simile of the vengeful lioness. Like Sarpedon's lion simile in Book 12 with its speculative outcome, this simile portrays a lion on the verge of attack. Both serve a similar dramatic function of introducing a *ritardando* at the moment before an encounter. The simile emphasizes the hero's return to the status of warrior at the moment he encounters his first opponent. His appearance now whets his appetite for vengeance and immortal glory, in contrast to his earlier refusal of mortal nourishment in Book 19. The lion is called σίντης, "ravening," an adjective used elsewhere in Homer of cattle marauders, the lion (11.481) and the wolf (16.353). It carries strong connotations of widespread, unlawful or sacrilegious destruction.[3] Like the participle ἀτίζων, σίντης has a special meaning for Achilles. It suggests on a much larger scale the consuming force within him to destroy animal and human life in a desperate attempt to gain revenge.

The simile intimates that the threat of the baneful beast is more ominous than usual. The countrymen all turn out to hunt down the beast. In the analysis of Agamemnon's *aristeia*, we saw that the juxtaposition of a simile with a digression can telescope two separate scenes into one more or less

---

[2] χόλος occurs elsewhere in an animal simile only at 22.94 (snake).

[3] In the *Odyssey* the verb σίνομαι is used of the Cyclopes plundering the Phaeacians (*Od.* 6.6); Teiresias warns Odysseus and his comrades not to feed on the cattle of the Sun (*Od.* 11.112). In post–Homeric Greek it means to pillage or waste a country, particularly the crops (Hes. *fr.* 117, Hdt. 5.74, cf. Theoc. 1.49).

homogeneous scene. Now in a similar fashion Achilles' simile complements two flashbacks in which Aineias and Achilles recall how the hero apprehended the Trojan prince on Mount Ida when he came after cattle, and how he sacked Lyrnessos and Pedasos (20.90ff., 188ff.). At the end of the abortive *Theomachy* Apollo singles out Aineias to fight against Achilles. The god inspires extraordinary strength into Aineias at 80 (μένος ἠύ), and because he is about to encounter such an ominous foe, breathes mighty power (μένος μέγα) into the "shepherd of the people" at 110. Aineias is thus chosen as defender of the Trojans and should be imagined in the simile as the herdsman leading the community in the attack against the marauder, just as in the narrative he attempts to lead an attack against Achilles, the destroyer of livestock. In the speech after the simile Achilles asks Aineias "Have the men of Troy promised you a piece of land, surpassing all others, fine ploughland and an orchard for you to administer to if you kill me?" His ironic question recalls the offer of the Aitolian leaders in the Meleager paradigm of the richest ground in Kalydon—half of it to be a vineyard and half of it unworked ploughland (9.577–80). Unlike Meleager, Achilles earlier refused the gifts promised him. Now, from the point of view of the Achaians, Achilles resembles Meleager returning to aid the Aitolians. From the Trojans' perspective, however, he continues to menace ravenously, as in the cattle–raids along the Troad.

In Books 21 and 22 there are several references to Achilles as a hunter. A simile at 21.573ff. likens him to the huntsman (ἀνδρὸς θηρητῆρος) who confronts the leopard emerging from a deep thicket (cf. Krischer's *"Standhalten"*). The use of the singular for "huntsman," like the pointed use of ἐλαφηβόλος to suggest Hektor in the lion simile at 18.318ff. strengthens the association. An earlier simile compares Achilles to an eagle, whose epithet is "the hunter," τοῦ θηρητῆρος (21.252). Toward the beginning of the Book he seizes twelve Trojan youths, compared to trembling fawns (21.29). As we saw earlier,[4] a more specific link between fawns and the Trojans at large establishes itself at the beginning of Book 22 when the Trojans, exhausted after the fighting in Book 21, are compared to terror–stricken fawns, πεφυζότες ἠύτε νεβροί (22.1). The participle πεφυζότες modifying the Trojans (21.6) recurs twice in the narrative towards the end of the Book (21.528, 532) and sustains the association with

---

[4] p. 38.

fawns.[5] The hunt metaphor aroused by this short comparison continues in the narrative when Achilles has his men bind the hands of the victims behind their backs with στρεπτοῖσι χιτῶσι and carry them to the ships. The action suggests a hunter binding the legs of the prey to facilitate transport, as Odysseus does with an impromptu rope when preparing to heave the stag planted by Circe (*Od.* 10.166–171). The Trojans are specifically chosen as a penalty, ποινή, to compensate for Patroklos' death (21.27f.). The action, and indeed the ritual, are reminiscent of the hunter's capture and the sacrifice of part of his kill to Artemis, who is mentioned in the Meleager paradigm and in Book 21.470ff.

The orderly manner in which Achilles goes about preparing for his revenge in Books 21 and 22 contrasts with the recklessness implied by his actions. References to Achilles as hunter create the illusion that his pursuit is sanctioned by codes of conduct, such as those that govern the hunt. Hunters capture, bind, and transport their prey; they erect trophies. But in choosing human victims Achilles has reversed the natural order, as Hektor had done earlier. The theme of illegal human possession with which the *Iliad* opens is now played out more violently in the animal metaphor. The lioness will gain revenge on the hunter for stealing her cubs. She will return to threaten the entire community in the guise of the marauder of livestock, just as Achilles now resumes his earlier plundering activities, but on a monumental scale. Although he focuses his anger on the murderer of his protégé, he consumes everything in his path. His revenge runs riot. But the actual manhunt does not begin until the next Book. Achilles' actions in Books 18 through 21 are only preparatory. Before Achilles can possess Hektor he must in a sense become him. He mimics Hektor's wish to treat his victim as a hunted trophy; the capture of the twelve youths imitates the killing of Patroklos. Poetically Achilles usurps Hektor's identification with the hunter by assuming temporarily his imagery.

## Imagery of Attack: Book 22

Book 22 is about heated pursuit and attack. The electrifying chase scene which forms the centerpiece resolves into an acrid encounter between

---

[5] Shipp, *Studies* 308, believes that πεφυζότες links the prologue of 21 to the last scene of the Book as well as to the opening of 22. The adjective φυζανικός occurs with deer at 13.102.

Achilles and his fatigued opponent. The death–scene has a precedent in Priam's empathetic vision, where he imagines dogs devouring his body in the forecourt of the besieged palace. Both scenes of attack are fraught with images of dogs: the hunting dog Sirios and his wild counterparts, the scavenger and, allusively, the rabid dog. As in Book 17 references to scavengers predominate, especially as verbal weapons in the confrontation scene. The similarity between the two attack scenes allows them to be taken together—as if out of Priam's dark imaginings Achilles' onslaught would emerge as a nightmarish reality.

The wide–ranging use of dogs in their various aspects to point up the degree of savagery in the warrior is evident in Book 22, in which table dog, hound, scavenger, and rabid dog converge. Living as closely as he does to man, the dog symbolizes a mortal's inability to renounce the savage instincts just beneath the skin that lead him to bloodthirsty fighting. Concurrently this inability to control impulses awakens in man the fear of degeneration to a savage state of which the logical conclusion is cannibalism. Folkloric beliefs about the dog in ancient Greece and other cultures express a concern that the wild instincts of the dog's ancestors, the jackal and especially the wolf, will resurface.[6] As the dog may revert to his scavenging ancestors, so man may lose touch with the restraints that distinguish him from the beast.

As Achilles is associated with the hunter, so he is linked with the dog as part of the reversal of roles between him and Hektor as hunter and hunted. Finally the leader of the Myrmidons becomes like a scavenger, and Hektor the animal's prey. In Book 22 Achilles is directly associated with the dog: once in a scavenging curse invoked by Priam (42f.), twice in similes, and again in a scavenging threat. The first simile (26ff.), comparing him to Orion's dog, the baleful dog–star Sirios, builds upon oblique references to Achilles as a plague–ridden, rabid dog at the end of the preceding Book; his coming generates ripples of fear about the eventuality of dogs

---

[6] These beliefs are encouraged by the phenomenon of the rabid dog, which is already attested in Homer by κύνα λυσσητῆρα (8.299). The belief that the wolf charms the dog back to its origins is evident in the word λύσσα, which in post–Homeric Greek means "rabies." It is derived from the radical of λύκος of which it is a feminine. S. Porzig, *Die Namen für Satzinhalte im Griechischen und im Indogermanischen, Untersuchungen z. indogerm. Sprach- u. Kulturwiss.*, 10 (1942) 349, believes that it signifies the demon, the she–wolf, that transforms the dog into a wolf; cf. B. Lincoln, "Homeric λύσσα: 'Wolfish Rage'," *IF* 80 (1975) 98–99; *contra* H. Erbse, *Untersuchungen zur Funktion der Götter im homerischen Epos, Untersuchungen zur antiken Literatur und Geschichte* 24 (1986) 36f.

feeding not only on Hektor's corpse, but on Priam's as well. The second simile, one of the eleven similes highlighting the chase scene, describes a hound pursuing a deer. Let us explore it in some detail, in relation to other similes in this scene.

Many of the similes in the chase scene deal with the swift movements of animals; three similes are accompanied by a formulaic line containing λαιψηρά and γόνυ. Others evoke various luminous qualities: the gleam of the dog-star; the fiery cold gaze of the snake; and finally, tranquil Hesperos. The similes are flanked by speeches. Descriptions of the internal thoughts of characters alternate with the movement of the chase, like a series of stills placed amid the larger, continuous action. The passage may be set out as follows:

| | |
|---|---|
| 22.15–20 | Achilles' speech to Apollo |
| 22f. | race–horse simile |
| 24 | ὡς 'Αχιλεὺς λαιψηρὰ ... καὶ γούνατ' ἐνώμα |
| 26–31 | dog–star simile |
| 38–89 | speeches of Priam and Hekabe |
| 93–95 | snake simile |
| 99–130 | Hektor's soliloquy |
| 139–142 | fire and hawk vs. dove similes |
| 144 | τεῖχος ..., λαιψηρὰ δὲ γούνατ' ἐνώμα |
| 162–64 | race–horse simile |
| 168–85 | Zeus' and Athene's speeches |
| 189–200 | dog vs. fawn and dream similes |
| 204 | ... ὅς οἱ ἐπῶρσε μένος λαιψηρά τε γοῦνα |
| 209–272 | scales; speeches between Achilles and Hektor |
| 297–30 | Hektor's speech |
| 308–318 | eagle vs. lamb or hare and Hesperos similes |
| 331–336 | Achilles' vaunting speech |

The momentum of the chase amid the static speeches is maintained by the action similes.[7] These involve animals and birds known for their swiftness: first, thoroughbred horses; and then pairs of animals, the hawk and dove, the eagle and lamb—animals who in nature eternally pursue or are pursued. In the realm of the hunt, dogs chase deer.

---

[7] Cf. J. Duban, "Distortion as a Poetic Device in the 'Pursuit of Hektor' and Related Events," *Aevum* 54 (1980) 3–22.

The theme of eternal pursuit also occurs in the famous dream simile, which forms a pair with the dog–fawn simile six lines earlier. The one restates in psychological terms what the other expresses through the hunting metaphor. In the dream simile both the pursuer and the pursued are interminably paralyzed in their tracks, adding to the frustration of the pursuer and the anxiety of the pursued. In the dog–fawn simile, the fawn takes refuge in a dense thicket, where she cowers in fear until the persistent dog tracks her down and flushes her from the copse. As in the dream, the outcome of this chase is indeterminate; the simile ends on an unresolved note, and we are left to imagine the race still in progress.

The dog–fawn simile has far–reaching significance for the overall hunting imagery. In the light of this simile let us reconsider that of the vengeful lioness at 18.318ff.

> Ἕκτορα δ' ἀσπερχὲς κλονέων ἔφεπ' ὠκὺς Ἀχιλλεύς.
> ὡς δ' ὅτε νεβρὸν ὄρεσφι κύων ἐλάφοιο δίηται        22.189
> ὄρσας ἐξ εὐνῆς διά τ' ἄγκεα καὶ διὰ βήσσας ·
> τὸν δ' εἴ πέρ τε λάθηισι καταπτήξας ὑπὸ θάμνωι,
> ἀλλά τ' ἀνιχνεύων θέει ἔμπεδον ὄφρά κεν εὕρηι ·
> ὡς Ἕκτωρ οὐ λῆθε ποδώκεα Πηλείωνα.

cf.

> τοῖσι δὲ Πηλείδης ἁδινοῦ ἐξῆρχε γόοιο        18.316
> χεῖρας ἐπ' ἀνδροφόνους θέμενος στήθεσσιν ἑταίρου
> πυκνὰ μάλα στενάχων ὥς τε λὶς ἠϋγένειος,
> ὧι ῥά θ' ὑπὸ σκύμνους ἐλαφηβόλος ἁρπάσηι ἀνὴρ
> ὕλης ἐκ πυκινῆς · ὃ δέ τ' ἄχνυται ὕστερος ἐλθών,        320
> πολλὰ δέ τ' ἄγκε' ἐπῆλθε μετ' ἀνέρος ἴχνι' ἐρευνῶν
> εἴ ποθεν ἐξεύροι · μάλα γὰρ δριμὺς χόλος αἱρεῖ ·
> ὡς ὃ βαρὺ στενάχων μετεφώνεε Μυρμιδόνεσσιν ·

In both similes a wild animal has been displaced from the lair. The enraged lioness seeks to avenge herself by sniffing out the tracks of the deerhunter; in the fawn simile the persistent hound tracks down the fawn by training and instinct. These two similes, so outwardly different, again reflect the reversal of Hektor's status from hunter to prey. Achilles changes from the lioness to a hunter with his hound, while Hektor changes from deerhunter to deer. The lioness, in effect, gains vengeance on him. Though no men are mentioned in the dog–fawn simile, it is, as we have seen in other examples, an excerpt from the hunt which focuses on the dog, while the men are suggested by metonymy. As Hektor is sometimes the hunter, sometimes the hound, Achilles is now the hunter's hound.

The following coincidences between the two similes may be observed. The details of tracking the victim (ἀνέρος ἴχνι' ἐρευνῶν and ἀνιχνεύων 18.321, cf. 22.192) are peculiar to those passages. ἴχνος and its derivatives do not recur in this sense in the *Iliad* (cf. *Od.* 17.317, *Od.* 19.436). The similes end inconclusively but ominously with a phrase in a conditional mood which expresses the desire and determination of the tracking animal to apprehend his prey: εἴ ποθεν ἐξεύροι (18.322), cf. ὄφρα κεν εὕρῃ (22.192). The locale of the pursuit in both similes is described as in "the folds" (ἄγκεα, 18.321, cf. 20.190) of the mountains.[8] The object of the hound's chase is the young of the deer, the fawn, called νεβρόν ... ἐλάφοιο. In the lion simile, and only here, is the type of hunter named.[9] Although the term is generic for hunter, "deerhunter" has a special ring to it: the pointed reference to Hektor as deerhunter corresponds to the reference to fawns in Book 21, which is carried on into the beginning of Book 22. In the animal metaphor the twelve Trojan youths are fawns, and Hektor the doe, as Achilles is the lioness and Patroklos the cubs. The lioness turned hunting dog gains exactly apposite vengeance on the hunter turned doe.

The bitterly cynical race–horse simile precedes the dog–fawn simile. There the poet dispels any illusion about the nature of the race: "Here was no festal beast, no ox–hide they strove for, for these are the prizes that are given men for their running. No, they ran for the life of Hektor, 'breaker of horses'." The race is so dreadful as to defy comparison. But the poet, using *praeteritio*, in fact compares the two men to race horses. The simile describes the gifts laid up for the chariot–race, either a tripod or a woman; the resumptive clause describes the sport it gives the gods (22.165f.). The mocking tone lingers in the dog–fawn simile. Achilles is still tracking his victim; but whereas before he was compared to a lioness justly angered at the theft of her young, he is now no more than a hound pursuing a frightened animal. As we will see, Achilles' metaphorical debasement in the animal world has not reached its lowest limit.

---

[8] The word recurs only twice in Homer: in a forest–fire simile (20.490) and in Menelaos' prophetic lion simile (*Od.* 4.335ff.).

[9] Leaf, *Commentary ad loc.*, quotes Eustathius: "ἡ διπλῆ ὅτι οὕτως τοὺς κυνηγοὺς καταχρηστικῶς λέγουσιν οἱ ποιηταί" and adds, "i.e. the passage shows that the word is not to be restricted to a hunter of deer."

At the beginning of Book 22 Priam has an apocalyptical vision, akin to a hallucination precipitated by the sight of Achilles. His fears spread to Hekabe and then to Andromache. Priam's vision is punctuated by the image of scavenging dogs. Amidst the destruction of the palace Priam sees his table dogs defile his body and lap up his blood. The violence of the language vividly evokes the fear hanging over the palace. Three indications suggest an association between Achilles and the rabid dog. In his frenzied command to the men guarding the gates Priam shouts a warning that Achilles is raging nearby, ἐγγὺς ὅδε κλονέων; he has a presentiment that a ruinous plague is to ensue and calls Achilles "the baneful man" (οὖλος ἀνήρ 21.533–36). In the narrative following Priam's command, Achilles' breast is said to be seized by a strong raging: λύσσα δέ οἱ κῆρ | αἰὲν ἔχε κρατερή (542f.).[10] The adjective οὔλιος, related to οὖλος and ὀλοός,[11] as in the phrase λύσσαν ὀλοήν, occurs in a star simile (11.62) comparing Hektor to the bright, baleful star; we may assume that it is Sirios. This series of associations indicates a possible link between Sirios and λύσσα.

When Priam sees Achilles for the second time, the warrior, blazing in all his armor, resembles the star that appears at the harvest. There is no doubt now that this is Sirios, the mightiest star in the heaven. Men also call it the "dog of Orion, ὅν τε κύν' Ὠρίωνος ἐπίκλησιν καλέουσι (22.29). The appearance of Sirios signals scorching heat; it was believed that in rising with the sun it added to the heat of the season. Although Sirios is the hunting dog of Orion, its traditional link with "the dog days" encourages an identification with the rabid dog. Pliny reports beliefs that the coming of the star made dogs go rabid.[12] In the present passage the star is said to

---

[10] Elsewhere οὖλος modifies λύσσα, used of Hektor, whose comparison to the rabid dog was noted above (p. 77, n. 16). In Homer λύσσα properly means "martial rage," but the etymology suggests that rabies is the primary meaning. It is used with Hektor and Achilles. In the Embassy Odysseus says that Hektor rages, μαίνεται ... κρατερὴ δέ ἑ λύσσα δέδυκεν (9.238f.), and he reiterates the point at the conclusion of his speech, by saying that he is seized by a terrible rage, λύσσαν ἔχων ὀλοήν (9.305). The short simile λυσσώδης φλογὶ εἴκελος (13.53) likening Hektor to a raging flame shows that the idea of rage is furthermore linked with fire.

[11] ὀλοός is used of Achilles at 24.39. S. Schein, *The Mortal Hero* (1984) 158, points out that only here is the adjective used of a person rather than a destructive force in nature.

[12] "Rabies canum sirio ardente homini pestifera, ut diximus, ita morsis letali aquae metu." Cf. "Canes quidem toto eo spatio maxime in rabiem agi non est dubium." Pliny, *NH* 8 152; 2 107; cf. Schol. in Arat. Phain. 27, 245, 7f. [ed. Maass]; Timotheus of Gaza, Περὶ Ζῴων, 26, 16, 24 and 31ff. [ed. Haupt].

bring πυρετός[13] to mortals: καί τε φέρει πολλὸν πυρετὸν δειλοῖσι βροτοῖσιν (22.31). The sight of the warrior engenders a desperate and morbid vision in the mind of King Priam:

πρὸς δ' ἐμὲ τὸν δύστηνον ἔτι φρονέοντ' ἐλέησον      22.59
δύσμορον, ὅν ῥα πατὴρ Κρονίδης ἐπὶ γήραος οὐδῶι
αἴσηι ἐν ἀργαλέηι φθίσει κακὰ πόλλ' ἐπιδόντα
υἷάς τ' ὀλλυμένους ἑλκηθείσας τε θύγατρας,
καὶ θαλάμους κεραϊζομένους, καὶ νήπια τέκνα
βαλλόμενα προτὶ γαίηι ἐν αἰνῆι δηϊοτῆτι,
ἑλκομένας τε νυοὺς ὀλοῇς ὑπὸ χερσὶν Ἀχαιῶν.      65
αὐτὸν δ' ἂν πύματόν με κύνες πρώτηισι θύρηισιν
ὠμησταὶ ἐρύουσιν, ἐπεί κέ τις ὀξέι χαλκῶι
τύψας ἠὲ βαλὼν ῥεθέων ἐκ θυμὸν ἕληται,
οὓς τρέφον ἐν μεγάροισι τραπεζῆας θυραωρούς,
οἵ κ' ἐμὸν αἷμα πιόντες ἀλύσσοντες περὶ θυμῶι      70
κείσοντ' ἐν προθύροισι. νέωι δέ τε πάντ' ἐπέοικεν
ἄρηϊ κταμένωι δεδαϊγμένωι ὀξέι χαλκῶι
κεῖσθαι· πάντα δὲ καλὰ θανόντι περ ὅττι φανήηι·
ἀλλ' ὅτε δὴ πολιόν τε κάρη πολιόν τε γένειον
αἰδῶ τ' αἰσχύνωσι κύνες κταμένοιο γέροντος,      75
τοῦτο δὴ οἴκτιστον πέλεται δειλοῖσι βροτοῖσιν.
Ἦ ῥ' ὃ γέρων, πολιὰς δ' ἄρ' ἀνὰ τρίχας ἕλκετο χερσὶ
τίλλων ἐκ κεφαλῆς· οὐδ' Ἕκτορι θυμὸν ἔπειθε.
μήτηρ δ' αὖθ' ἑτέρωθεν ὀδύρετο δάκρυ χέουσα
κόλπον ἀνιεμένη, ἑτέρηφι δὲ μαζὸν ἀνέσχε·      80
καί μιν δάκρυ χέουσ' ἔπεα πτερόεντα προσηύδα·
Ἕκτορ τέκνον ἐμὸν τάδε τ' αἴδεο καί μ' ἐλέησον
αὐτήν, εἴ ποτέ τοι λαθικηδέα μαζὸν ἐπέσχον·
τῶν μνῆσαι φίλε τέκνον ἄμυνε δὲ δήιον ἄνδρα
τείχεος ἐντὸς ἐών, μὴ δὲ πρόμος ἵστασο τούτωι      85
κλαύσομαι ἐν λεχέεσσι φίλον θάλος, ὃν τέκον αὐτή,
οὐδ' ἄλοχος πολύδωρος· ἄνευθε δέ σε μέγα νῶιν
Ἀργείων παρὰ νηυσὶ κύνες ταχέες κατέδονται.

Priam imagines his sons slain in the toils of war, his daughters and daughters–in–law dragged away captive, the chambers of the palace ravaged, and the innocent children dashed to the ground. As for himself the

---

[13] A *hapax legomenon* in Homer; in later Greek it means "fever."

king fears that the very dogs he raised, those who attend him at the feast and watch the gates (τραπεζῆας θυραωρούς), will turn into wild, flesh–eaters (ὠμησταί): they will maim his body and lap up his blood. The flesh–eating dogs (66f.) are anticipated a few lines earlier by other images of scavenging dogs, suggested by the participles ἑλκηθείσας (62) and ἑλκομένας (65) from ἕλκω, used of dogs pulling and tearing away in a simile at 17.558, and later in this Book at 335f. in a scavenging threat to Hektor.[14] Priam's fear is contagious. Hekabe entreats Hektor from afar to desist from battle. His death will mean for her and Andromache the deprivation of the very right of relatives to mourn his death properly, since Hektor will be far away from the ships of the Argives where the running dogs will feed on him.[15] Later in the Book Andromache rephrases Hekabe's words (22.508–14). Hektor now lies dead. She imagines his naked corpse lying far away and being devoured by the dogs and birds. In desperation she decides to perform a mock funeral ritual by burning his clothing.

Priam's obsessive fear invades him at the moment he realizes that his favorite son, the protector of the city, will fall. No worthy heir will succeed him. After Hektor's death, and in the absence of any worthy successors to perpetuate his stock, the dogs in the vision in effect become the inheritors. The last detail of Priam's feverish vision describes the dogs defiling his grey head and beard and ripping apart his genitals—a symbolic quenching of his power.

What is a vision for Priam borders on reality for Hektor as the threat to the scavenger rises prominently to the surface. In his vision Priam empathetically experiences his son's death. The theme of vulnerability suggested in Priam's speech carries on in the following narrative. The exposure of the body, as we have seen in Book 17 when Hektor denuded Patroklos' corpse of its armor, lays it open to the attack of the scavenger. Clothing and armor are peculiar to men and are one of the marks distinguishing men from animals. An animal may have its own form of natural protection—a thick hide or a pair of tusks; a man has no natural bodily protection. When deprived of his artificial defences, his vulnerability increases, and he loses the accoutrement that sets him apart from all other creatures. Worse yet, a denuded corpse sinks to the level of carrion.

---

[14] The participle κεραϊζόμενος (63) is a formulaic word in similes describing the marauding lion (5.557 and 16.752).

[15] See Segal, *Theme* 62, for the added emphasis of μέγα in the otherwise formalized phrase: ἄνευθε δέ σε μέγα νῶϊν (88).

Rape, here a metaphor for vulnerability, lies behind the outrages against his daughters and daughters–in–law, as well against the marriage–chambers. The attack on the palace of Priam amounts to a violent penetration. Priam himself is impaled on the sharp bronze (67f.). The flesh–eating dogs too pierce him with their sharp teeth and defile his genitals. In keeping with the metaphor of the vulnerability of the city, the passage emphasizes doors.[16] Priam stands at the threshold of old age (60); the dogs who formerly stood watch at the doors (θυραωρούς, 69) of the palace now attack and devour Priam at the front doors of the palace (πρώτῃσι θύρῃσιν, 66); they lick his blood from the body lying in the forecourt (προθύροισι, 71). The dogs behave like scavengers. The doors separate the area inhabited by humans from that ἀγροῦ ἐπ' ἐσχατιήν, beyond which the scavenger normally lives.

The metaphor of vulnerability anticipates Hektor's soliloquy, where he expresses the fear that Achilles will not pity him but will slay him, "naked as a woman" (κτενέει δέ με γυμνὸν ἐόντα | αὔτως ὥς τε γυναῖκα, 124) once he has stripped off the armor. Here the interchangeability between death, vulnerability, and rape is more succinctly expressed. At the end of the Book Andromache imagines the body lying naked, eaten by worms, once the dogs have sated themselves (22.508–10):

> νῦν δὲ σὲ μὲν παρὰ νηυσὶ κορωνίσι νόσφι τοκήων,
> αἰόλαι εὐλαὶ ἔδονται, ἐπεί κε κύνες κορέσωνται,
> γυμνόν· ...

In the actual confrontation between Achilles and Hektor he eyes the Trojan's beautiful flesh to discern "where it would give the biggest opening" for his spear, εἰσορόων χρόα καλόν, ὅπῃ εἴξειε μάλιστα (321). From the armor that Patroklos once wore emerges Hektor's throat, "where the collar–bones hold the neck from the shoulders, where death of the soul comes most quickly." Achilles hurls the spear and penetrates the neck. The passage at 317–330 deals with exposure and vulnerability, epitomized by the chink in Hektor's armor. In the supplication scene there are four references to scavenging dogs (335, 339, 348, 354), and one term of abuse (κύον, 345) in close succession, all of which serve to intensify the earlier image of the scavenger. In this climactic scene the various associations

---

between Achilles and the symbols of destruction—fire, star, dog–star, scavenging dog, and spear—brilliantly harmonize.

> οἷος δ' ἀστὴρ εἶσι μετ' ἀστράσι νυκτὸς ἀμολγῶι        22.317
> ἕσπερος, ὃς κάλλιστος ἐν οὐρανῶι ἵσταται ἀστήρ,
> ὣς αἰχμῆς ἀπέλαμπ' εὐήκεος, ἣν ἄρ' Ἀχιλλεὺς
> πάλλεν δεξιτερῆι φρονέων κακὸν Ἕκτορι δίωι        320
> εἰσορόων χρόα καλόν, ὅπηι εἴξειε μάλιστα.
> τοῦ δὲ καὶ ἄλλο τόσον μὲν ἔχε χρόα χάλκεα τεύχεα
> καλά, τὰ Πατρόκλοιο βίην ἐνάριξε κατακτάς·
> φαίνετο δ' ἧι κληῖδες ἀπ' ὤμων αὐχέν' ἔχουσι
> λαυκανίην, ἵνα τε ψυχῆς ὤκιστος ὄλεθρος·        325
> τῆι ῥ' ἐπὶ οἷ μεμαῶτ' ἔλασ' ἔγχεϊ δῖος Ἀχιλλεύς,
> ἀντικρὺ δ' ἁπαλοῖο δι' αὐχένος ἤλυθ' ἀκωκή·
> οὐδ' ἄρ' ἀπ' ἀσφάραγον μελίη τάμε χαλκοβάρεια,
> ὄφρά τί μιν προτιείποι ἀμειβόμενος ἐπέεσσιν.
> ἤριπε δ' ἐν κονίηις· ὃ δ' ἐπεύξατο δῖος Ἀχιλλεύς·        330
> Ἕκτορ ἀτάρ που ἔφης Πατροκλῆ' ἐξεναρίζων
> σῶς ἔσσεσθ', ἐμὲ δ' οὐδὲν ὀπίζεο νόσφιν ἐόντα
> νήπιε· τοῖο δ' ἄνευθεν ἀοσσητὴρ μέγ' ἀμείνων
> νηυσὶν ἔπι γλαφυρῆισιν ἐγὼ μετόπισθε λελείμμην,
> ὅς τοι γούνατ' ἔλυσα· σὲ μὲν κύνες ἠδ' οἰωνοὶ        335
> ἑλκήσουσ' ἀικῶς, τὸν δὲ κτεριοῦσιν Ἀχαιοί.

There are several details in this passage which suggest the latent presence of the scavenger. The Hesperos simile, here a paradoxically tranquil comparison, looks back to the earlier comparison between Achilles and the deadly dog–star. The passage distinguishes sharply between the artificial protection of the armor and the vulnerable flesh beneath: χρόα καλόν … χρόα χάλκεα τεύχεα, | καλά … (322f.). Like a scavenger, Achilles with his spear lunges for the flesh which he had greedily eyed. ἀκωκή, the word for spear–point here as on other occasions, is semi–animate and suggests the fangs of the scavenger.[17] The epexegetical phrase ἧι κληῖδες ἀπ' ὤμων αὐχέν' ἔχουσι (324) pointing to the vulnerable area where the spear enters again emphasizes the penetrability of the flesh (sc. 321f.).[18] As

---

[17] Cf. 20.260, 21.60. 22.327=17.49 (death of Euphorbos) followed by stripping of armor.

[18] In a similar usage at 8.325f. (cf. 5.146), Hektor wounds a warrior in the neck ὅθι κληΐς ἀποέργει, and this is said to be the most fatal of all places, μάλιστα δὲ καίριόν ἐστι. In the present passage a similar exegesis is given: ἵνα τε ψυχῆς ὤκιστος ὄλεθρος

Hektor lies dying the skin covering his body is no longer referred to as "flesh" (χρόα 321) but as "meat" (κρέα 347). The aggressor morbidly desires to eat him raw (ὤμ' ἀποταμνόμενον κρέα ἔδμεναι ... 347) like a scavenger. Achilles' wish implies the denial of the fundamental distinctions between the raw and the cooked. It amounts to a perversion of the rights of the lead warrior to be honored by meat and wine celebrated in Sarpedon's *parainesis* (12.310ff.). But perhaps we should not be surprised that Achilles has become so estranged from the human sphere; he has already nullified two institutions that separate man from all other creatures by refusing to exchange oaths and by failing to acknowledge the claims of a suppliant.

In the preceding scene Hektor promises to relinquish Achilles' corpse should he slay him but seeks, in vain, to exact an oath from Achilles to reciprocate. Achilles only scoffs at him:

Ἕκτορ μή μοι ἄλαστε συνημοσύνας ἀγόρευε·      22.261
ὡς οὐκ ἔστι λέουσι καὶ ἀνδράσιν ὅρκια πιστά,
οὐδὲ λύκοι τε καὶ ἄρνες ὁμόφρονα θυμὸν ἔχουσιν,
ἀλλὰ κακὰ φρονέουσι διαμπερὲς ἀλλήλοισιν,
ὡς οὐκ ἔστ' ἐμὲ καὶ σὲ φιλήμεναι, οὐδέ τι νῶϊν      265
ὅρκια ἔσσονται, πρίν γ' ἢ ἕτερόν γε πεσόντα
αἵματος ἆσαι Ἄρηα ταλαύρινον πολεμιστήν.

This parable, in the form of a simile, is comparable to the αἶνος of the hawk and nightingale which Hesiod (*Erga* 202ff.) uses to instruct Perses about the law of the stronger.[19] Achilles' present use of the parable is also didactic as he usurps one of the poet's most highly developed compositional devices. But he refuses to relate to Hektor on human terms; he fails utterly to acknowledge the concept of reciprocity so fundamental to heroic ideals. Achilles' blending of the human and animal violates the distinction in pre–Classical Greece between man and beast, as defined, for example, by Hesiod (*Erga* 274–285): Zeus has apportioned the law of δίκη to men alone; animals are left to feed upon one another "since there is no δίκη among them."

---

(22.325, cf. ὅπη εἴξειε μάλιστα, 321). In Achilles' famous speech in the Embassy the ψυχή is not λεϊστή once it has passed this barrier (9.408f.).

[19] On the earlier hawk simile (22.139-42) and its relation to the parable and the Hesiodic passage see S. H. Lonsdale, *Hermes* 117 (1989) 408-409.

The final exchange of words between Hektor and Achilles is a bald restatement of their earlier interchange. Achilles tells him that the dogs and vultures will foully rip him apart (22.335f.). Hektor, whose strength is dwindling, drops to his knees and supplicates the heartless warrior. He presents only one simple demand, expressing it succinctly in a single hexameter verse: μή με ἔα παρὰ νηυσὶ κύνας καταδάψαι Ἀχαιῶν (339). Achilles replies, bitterly calling Hektor "a dog" (345). But it is the former, who, in effect, becomes the scavenger when he reveals his desire to feed upon the flesh: "I wish only that my spirit and fury drove me | to hack away your meat and eat it raw..." (346f.) A scavenging threat in the next verse (348) punctuates his sentiments. At the end of his speech Achilles promises to carry out what Hekabe most feared: to deny the mother the chance to mourn her child: "... not even so shall the lady your mother | who herself bore you lay you on the death–bed and mourn you: | no but the dogs and the birds will have you all for their feasting." (352–54) Achilles extracts his spear from the flesh, and the rest of the Achaians come running about Hektor. They crowd around the corpse and prick the flesh with their spears—ignoble warriors who have had no part in the kill. They say with irony: "See now, Hektor is much softer to handle than he was when he set the ships ablaze with the burning firebrand" (373f.).

Fears and uncertainty about the status of Hektor's corpse continue to smoulder until Priam actually pays the ransom for Hektor.[20] We learn, however, much earlier that the fate worse than death was not to befall Hektor. At 23.182f., after piling high Patroklos' pyre, Achilles utters an imprecation to the corpse for the last time: Ἕκτορα δ' οὔ τι | δώσω Πριαμίδην πυρὶ δαπτέμεν, ἀλλὰ κύνεσσιν. With characteristically abrupt Homeric antithesis we learn in the next succinct line that the dogs did not "busy around the corpse of Hector," ... ὣς φάτ' ἀπειλήσας· τὸν δ' οὐ κύνες ἀμφεπένοντο (23.184). After the slaying, Aphrodite keeps the dogs away night and day, ministering to the corpse. She bathes it with ambrosial oil, while Apollo shrouds it with a dark cloud to keep away the bite of the sun. The verb ἀμφιπονέομαι interacts with either the theme of attendance or with the rituals of feasting and the funeral, as in the present passage, which occurs immediately after Patroklos' funeral.[21] Here this

---

[20] The scavenger theme recurs at 23.21, 183–85, 24.211, 409, 411.

[21] The Paphlagonians, for example, care for the body of their fallen comrade and carry him back to sacred Ilion: τὸν μὲν Παφλαγόνες μεγαλήτορες ἀμφεπένοντο (13.656). After Machaon has ministered to the wounded Menelaos, the Achaians attend

reassuring phrase cushions us from Achilles' estranged vision and realigns us with a more charitable set of values, whereby the gods take pity on mortals. We are reminded of a more familiar and ordered world, where reciprocity exists among men in the feast and in the funeral, and where the table dogs feast with their masters rather than devour them.

---

to him, ὄφρα τοὶ ἀμφεπένοντο βοὴν ἀγαθὸν Μενέλαον (4.220). In the *Odyssey* the often thin line of distinction between the wedding feast and the funeral emerges in Odysseus' reproving speech to Ktesippos, "I would have struck you with my spear, and instead of your marriage your father would have been busy with your funeral here" (*Od.* 20.307). In Eumaios' homeland the men feasting with his father attend to him: ἀνδρῶν δαιτυμόνων, οἵ μευ πατέρ' ἀμφεπένοντο (*Od.* 15.467).

# Chapter VII
## Conclusions: Animal Imagery in the Homeric Narrative

Ancient Near Eastern literature indicates that similes had been a literary device at least a millennium before the Homeric poems assumed their final shape. In Greek as in Mesopotamian oral poetry the animal was one of the earliest and most durable simile types, along with similes of fire and other natural phenomena. The success and frequency of the animal simile as a type with distinctive emotional and psychological overtones depended on its familiar and universal subject matter, drawn as it was from the pastoral and hunting experiences of eastern Mediterranean audiences.

Can the same first–hand experience be claimed for the scene sketched in the lion simile, the master simile and the most frequent type in Homer? Although the lion simile in part owes its prevalence to the same sort of motivation that induced rival singers to parade their virtuosic knowledge of horse–lore, it is uncertain if lions themselves were prevalent as living models in Asia Minor toward the end of the oral tradition. The apparent popularity of lion imagery in the *Iliad* can be explained by two lines of argument, which are more or less mutually exclusive. Either the threat of the marauding lion to livestock could be related, like other simile subjects, to the everyday experiences of Ionian audiences. Or else the audience, totally unfamiliar with the lion as an indigenous species, enjoyed hearing about an exotic creature imported from a Near Eastern context. According to the latter argument, lion similes would then have added an "oriental" flair to the epic. Obviously these alternatives could be resolved if we had conclusive proof that lions were a widespread species in Archaic Ionia, which cannot be automatically assumed.[1] Unfortunately, the evidence for lions in the

---

[1] Our knowledge about the presence of lions in the Greek world at large is confined to the Bronze Age and to the classical period. It is widely assumed that lions were present in Anatolia; it is unclear when, if ever in antiquity, they disappeared from the coast of Asia Minor. The lack of evidence for lions in the intervening Archaic period does not necessarily mean that lions became extinct. Herodotus (7.125–27) reports lions attacking Xerxes' pack–camels in Thrace in the early fifth c., BC. The detail of the lions'

Archaic period is limited to the epic and to the frequent representation of lions in Archaic art.[2] In the absence of objective written testimony or osteological evidence,[3] one argument in favor of the audience's first–hand knowledge of the lion is the interchangeability of the lion with the boar in

---

curious penchant for the malodorous pack–camels (while they left untouched the other pack–animals and men) supports the general impression of a reliable eye–witness account. He limits the geographical distribution of the lion to the area between the Achelous River in Acarnania and the Nestus, which flows through Abdera, and comments that there are many lions living in that region. Aristotle (*HA* 606$^b$1.14ff.) also names the Achelous and Nestus rivers as boundaries for the distribution of the lion in Thrace, but says that lions there are rare. Aristotle may have taken his information from Herodotus, but it is unlikely that he followed his account blindly without foundation. As a native of Stagirus and resident in Macedonia, Aristotle presumably would have had a chance to verify, and if necessary rectify, Herodotus' testimony. Aristotle elsewhere corrects Herodotus where he strays on points of natural history, such as his misinformed account of the parturition in the lioness (*HA* 579$^b$ 3f., cf. Hdt. 3.108). Cf. A. B. Meyer, "The Antiquity of Lions in Ancient Greece," *Annual Report of the Board of Regents*, Smithsonian Institution (1903) 661–67. For later versions of Herodotus' account, cf. Paus. 9.21.2, Ael., *NA* 18.36; Pliny, *NH* 8.45. For the distribution of the lion in the Mediterranean world, see S. Gsell, *Histoire ancienne de l'Afrique du Nord* (1913), 1, iii, n.2. Körner, *Homerische Tierwelt*, 9, 14, believes that the lion was a well–known phenomenon in Asia Minor contemporary with the performance of the poems. "Das Leben des Löwen ist bis in die Mitte des 19. Jahrhunderts nirgends vollständiger und richtiger beschrieben worden als in der Ilias ... Die Genauigkeit fast aller Schilderungen aus dem Leben des Löwen zeigt, daß der Löwe zur Zeit, als die Ilias an der kleinasiatischen Westküste gedichtet wurde, dort haüfig und wohlbekannt gewesen ist." The available evidence for sightings of lions in the Near East in recent times is summarized in D. L. Harrison, *The Mammals of Asia Minor* (1972) 3, 621ff.; cf. C. Danford and E. Alston, "On the Mammals of Asia Minor, Part II," *Proc. Zool. Soc. Lond.* (1880) 50–66.

[2] G. E. Markoe, "The 'Lion Attack' in Archaic Greek Art: Heroic Triumph," *CA* 8 (1989) 868–92, argues that the widespread representation of lions in Greek art of the late eight and seventh centuries is directly influenced by epic poetry. On the lion in Mycenaean art, see G. Mylonas, *AAA* 3 (1970) 421–25.

[3] Osteologically lions are attested in the Aegean Bronze Age. At Keos (Ayia Irini) the carnassial teeth from two lions were found. Neither was pierced for stringing, but they were called by J. L. Caskey "souvenirs by someone who had been in lion country." (Correspondence with the excavator.) At Tiryns the heel–bone of a small lion (*Panthera leo*) was discovered in the west wall near the skeleton of a man and dated by stratigraphy to c. 1230 BC. It was originally considered a talisman until the subsequent discovery in 1980 of another lion bone (part of the humerus) in a stratum containing pottery from the Shaft Grave period (16th c., BC). The bone was cut off at the shank, as is usually the case with oxen, sheep, goats, pigs, red deer, and dogs. The conclusion that the lion had been slaughtered for food is not altogether impossible. Lion bones with characteristic traces of cutting are known from the same period at Bogazköy. (J. Boessneck and A. von den Driesch, "Ein Löwenknochenfund aus Tiryns," *AA* 94 (1979) 447–49, cf. *AA* 96 (1981) 257f.) Boessneck and von den Driesch have subsequently located at least two bones at Kalapodi in Central Greece. (Correspondence with J. Boessneck.) Leopard bones were discovered in Troy VIIa and VIII (C. W. Blegen, *Troy* [1958] IV.1, 51; cf. 117, 123, 263).

similes. No one has claimed that the Greeks were unfamiliar with this creature. Moreover, if lions were known only from Near Eastern sources, as is sometimes argued,[4] we should expect the lion to appear only in short similes or epithets such as θυμολέων, with little more resonance than βοῶπις or γλαυκῶπις. Instead we find full–blown similes, with a convincing degree of realism that can also be found in the lion vignette on the Shield.

On the Shield, as in the majority of the lion similes, the lion is a marauder who jumps an enclosure to feed on the farmer's flocks within. As such he is a parasite, a noxious pest, a virtual scavenger.[5] Once in a simile at 11.474ff. the lion chases away jackals feeding on the carrion of deer so that he may himself devour the beast. Aristotle, in the very passage in the *Historia Animalium* (629 27ᵇff.) in which he admires Homer for his accurate portrayal of lions' fear of fire (cf. 11.553), remarks on the phenomenon of lions who feed on the farmer's livestock. He says that they are aging lions. "They invade the cattle–folds and attack human beings when they grow old and so by reason of old age and the diseased condition of their teeth are unable to pursue their wonted prey." Whatever the zoological reality behind the Homeric lion, it is clear that the symbol of the kingly, quasi–divine, lion from the Near East has been recreated in the similes in the image of the Homeric hero with all his strengths and weaknesses, but without sacrifice of naturalism. The variousness of the lion's behavior can then be explained not only by the shifting demands of the narrative context, but also by a faithful portrayal by the epic tradition of the smaller feline presumably in question, *Leo leo persicus,* known in Asia Minor in historical times.[6] It is possibly this more compact (στρογγυλώτερος) lion that

---

[4] Richter, *Landwirtschaft* 38; cf. T. B. L. Webster, *From Mycenae to Homer* (1958) 222; *contra* Steier, *RE* 13.986; J. K. Anderson, *Hunting in the Ancient World* (1985) 15.

[5] Cf. 3.23ff., 13.198ff. and M. Mueller, *The Iliad* (1984) 118, for the scene of a lion and a carcass. G. Guggisberg, *Simba, the Life of the Lion* (1961) 122, reports that once a pride of lions discovers domesticated animals to feed on they rarely revert to tracking down wild animals, an activity which requires a greater expenditure of energy than preying on sedentary animals.

[6] The Persian lion, *Leo leo persicus,* was officially identified by Fischer, cf. A. B. Meyer, *Dissertatio inauguralis anatomica–medica de genere felium* (1826) 6. The ambivalence of the lion's character is a known fact. In a modern study of the African lion in a pastoral context, G. Guggisberg, *op. cit.* 62, writes, "lazy as lions normally are, they can, on occasion be either extremely bold or astonishingly timid... A piece of cloth fluttering in the wind can frighten him away... It is possible to protect horses, mules, and oxen by simply hanging up pieces of cotton."

Aristotle (*HA* 629ᵇ33–630ᵃ1) had in mind when contrasting it with the larger, more courageous variety.

The Homeric lion simile *par excellence* consists of the feline attacking cattle, while the farmer and his dogs defend the herd. This dominant type then admits variations to suit virtually every narrative situation from attack to flight in battle episodes. Alongside the brave, attacking lion, we find the cowardly lion in similes that mark the retreat of a warrior. A lion simile is capable of being shaped to describe a purely psychological or even unconscious state, such as Penelope's fitful sleep at *Od.* 4.791f.

The basic scenario of men and dogs confronting a wild animal in the lion similes recurs in hunting similes. The alternative simile subjects "lion or boar" indicate that to a certain degree herding and hunting similes were interchangeable. We even find examples of the fusing of developed lion and boar similes. Lion and boar similes existed independently, of course, but the human and animal actors, settings, and actions they share make them so similar in essentials that the smaller group of hunting similes should be regarded as a series of variations on the dominant lion type, corresponding to different narrative situations. Although lion similes alternate between descriptions of locomotion and emotion, they tend to be more psychological in emphasis than hunting similes, which stress action and stubborn defense. Unlike the lion, which never appears outside of the similes and the Shield as a part of the plot, the boar is the subject of two memorable digressions of considerable length.

The highly successful poetic device of the epic simile in the Homeric text may be compared with other efforts in early Greek hexameter poetry. Although the surviving works and fragments of Hesiod do not yield a sufficient number of developed similes to draw a comparison,[7] the duel of Kyknos and Herakles in the pseudo–Hesiodic *Aspis* (367–486) contains eight developed similes, four of which form a simile chain (*Sc.* 386–412). Except for a simile at *Sc.* 42–45[8], all *Aspis* similes appear in the duel scene. A substantial portion of the rest of the poem of 480 lines comprises the ecphrasis of Herakles' shield (141–317) and direct speeches. The proportion of simile to narrative lines in the duel scene is 55:113 (45%). The density of similes far exceeds the highest concentration of similes in the

---

[7] The only developed simile in Hesiod is the fine bee simile (*Th.* 594–599, cf. *Erga* 304–6); cf. the short dog simile (*Th.* 834).

[8] These lines are rejected by Mazon.

dual scene in *Iliad* 22, which, as seen above,[9] contains seven similes, accounting for only 28 out of 404 lines (10%).

The subjects of the similes in the *Aspis* include falling boulders and trees, boar, grasshopper, a pair of lions in the wild, vultures, and a marauding lion. The last resembles our model Homeric simile (20.164ff.), especially in the detail of the beast lashing the tail against the rib–cage (*Sc.* 430f., cf. 20.170f.) Again the lion's anger wells up within (*Sc.* 429f., cf. 20.169), while outwardly the feline glowers (γλαυκιόων, *Sc.* 430, cf. 20.172). But one detail is new. The lion attacks with strong claws (ὀνύχεσσι), a word used only of birds' talons in Homer.

As in Homer, one finds in the *Aspis* alternate simile subjects, examples of ring composition, and images of striking intensity.[10] The boar simile (*Sc.* 387ff.) presents the familiar Homeric situation of the resolute beast confronting his attackers. It sharpens the image by describing the physical manifestation of the boar's anger: he whets his shining tusks, his mouth foams, his eyes glare. There is no attempt to balance the outward physical description of the beast with allusions to its inner courage and resolution. The narrative situation with which the boar simile is juxtaposed has Herakles jumping out of his chariot, a less congruous bridging of simile and narrative than we should expect in Homer, especially since Herakles has already descended from his vehicle at 370. In Homer the principle of using a simile that parallels or at least suits the narrative context is rarely violated. The avalanche simile (*Sc.* 374ff.) provides another example of a simile that does not altogether sit comfortably in its context. It describes a dramatically plausible scene from nature: cascading boulders crush the trees in their path. Such a simile would be found to be quite apposite in Homer if it described one army overpowering another, but it would be out of place in a duel scene, as it is here.

The *Aspis* similes differ from their Homeric counterparts in form and tone. The resumptive clauses are often omitted in favor of moving on to a new and seemingly unrelated simile. In the simile chain (*Sc.* 387–412) the subjects skip from boar, to grasshopper, to a pair of lions, to a pair of vultures, with a single resumptive clause at the end that summarizes only the last vulture simile. The only unifying thread is the animal subject. The chain

---

[9] Pp. 92–102.

[10] Alternate subjects: e.g., fire and whirlwind, 345; ring composition, e.g., ἐπ' ἀλλήλης δὲ πέσωσι (375) ... ὡς οἵ ἐπ' ἀλλήλοισι πέσον (379).

lacks the coherent development of the pastoral similes before the Catalogue of Ships (2.455–483) in which every member of the chain has a resumptive clause that effectively links together the passage. The grasshopper simile (*Sc.* 393–400) attempts to contrast the military subject of the poem with a bucolic image, but fails in its quaintness. In a heroic poem it is perhaps inappropriate to mention that the grasshopper's sustenance is the delicate (θῆλυς, *Sc.* 395) dew, and irrelevant to introduce two lengthy temporal clauses (*Sc.* 397–400) that dwell on the summer season in which the insect sings. Details which in a Homeric simile would somehow arrest or charm are extraneous here. The impression of haphazard composition in the *Aspis* similes is reinforced by comparing the ecphrastic description of the Shield of Achilles with that of the Shield of Herakles. We find gods and heroes interspersed with vignettes drawn from nature and contemporary society. As in the simile chain, there is a surfeit of subjects. There is, moreover, no careful balancing of scenes, as one finds, for example, in the two ambush scenes (one with human, the other with animal subjects) at 18.519–40 and 18.573–86. Instead, we are confronted with an agglomeration of subjects, with occasional attempts to integrate the human and animal, as for instance in the strained usage ὁμιληδὸν στίχες (*Sc.* 170) for the implausible situation of antithetical ranks of boars battling lions.

The foregoing comparison indicates that there are at least three ways to judge the quality of a simile: first by its internal excellence as an independent "poem," second by its integration into the narrative, and third by its relationship with other imagery. If we look retrospectively at the many herding and hunting similes, we can observe that the Homeric oral tradition adhered sufficiently to the basic model to maintain the existence of a type that allows individual similes to reverse or echo one another. At the same time the singer has varied the internal pictorial content by adding or suppressing details in order to avoid monotonous stereotypes. Each simile has an aura of spontaneity and freshness about it, even though created from a basic type. The poet neither adheres to his type too closely nor diverges fancifully from it, except in occasional similes noted above that distort naturalism to suit the narrative context (e.g., 13.200). The judicious variation of internal detail allows for dovetailing with the narrative. In the stitching of simile and its context the *Iliad* poet was a consummate master. He brought about integration in small details, for example by repeating verbs of motion and emotive terms, by introducing close parallelism or striking contrast of narrative and simile actors, or by adjusting the scale and tenor of the simile to match those of the narrative. Thus when Achilles first returns to the fighting we find a monumental lion simile that amplifies the lion simile type by joining apt and significant details. The singer crafted similes to narrative in more fundamental ways by reproducing on a miniature scale the rhythms of speech, with its emphasis on emotion, and those of narrative, with its objective voice and emphasis on action. The developed simile thus has an internal narrative in addition to its pictorial force. The alternation of points of view between wild beast on the one hand, and man and his domesticated animals, on the other, mirrors the poet's ideal objectivity in battle narrative that has him modulate from describing the battle from the point of view of the Achaians to that of the Trojans. From a thematic point of view, the destructive activities of the lion in similes condense cattle raids found in digressions, and those of the boar reproduce on a miniature scale aspects of the Meleager paradigm. The singer adapts the motifs of theft and vengeful destruction to the moral dilemmas in the poem: what rightfully belongs to a man, and what should he do when denied his due portion? In the similes the loss of property is felt by the farmer or shepherd, who, in contrast to the sovereign hero, represents the common man. This everyman is the epic audience.

The chief feature that characterizes the *Iliad* similes as a whole is the unity imposed on the imagery. The herding and hunting similes are linked not so much by their animal subject as by the inimical relationship between man and nature which they evoke. The hostility between man and beast is part of a larger theme setting nature against people, and which is epitomized in the storm, river, and sea similes. Although the mortals in the similes show an intimacy with the natural world, supernatural forces behind the scenes sometimes overwhelm and destroy them. In an epic that stresses the violence of action and the perils of hesitation, the singer adopts the herding and hunting similes to portray the constant strife that in time of war pits man against man. As is evident from the *Aspis* similes, in the hands of a less than consummate singer the similes are used at times inappropriately with insufficient regard for their subjects of comparison or context. The composer of the *Iliad* similes, however, has restricted the *agon* in such a way that the Achaian aggressors tend to be comparable to wild beasts, while the Trojan defenders are likened to the men and dogs. The restriction does not apply to Hektor; it is as though the poet wanted to emphasize that his forces could not be contained by any categorization, even a metaphorical one. When Hektor overtakes Patroklos he also takes over his image of the attacking lion. When Achilles returns to the fighting he is compared to the lion, but as his vengeance impels him to consume Hektor he also subsumes the Trojan's dominant imagery of the hunt. The singer charts and mirrors the fortunes of Hektor by establishing, then reversing, the hunting imagery. Achilles in a sense borrows Hektor's imagery, much as Hektor wears Achilles' fatal armor.

## Interaction of Animal Simile and Other Contexts

In reviewing the congruence of animal similes with narrative we may also explore the interaction of this poetic device with specific contexts, including non–animal similes, omens, epiphanies, ecphrastic descriptions, paradigms, and the scavenging motif. In the case of omens and epiphanies their close formal affinities with similes invite speculation about the possible supernatural dimensions of similes. Close correspondences between similes and other contexts at the levels of form, content, and diction carry over into theme and function. The motifs of destruction and vengeance in the marauding lion and boar similes are ultimately related to the pretexts for war in epics that account for the origins of human conflicts in disputes over animals stolen or divinely dispatched.

Much of the work on similes has contributed to our understanding of the different types by which they can be grouped. But the common phenomenon of alternative subjects within a simile indicates that subjects were to a certain extent interchangeable. Usually the alternatives do not diverge appreciably from one another. The lion is bracketed with the boar (e.g., 8.338), or different varieties of trees are given (e.g., 16.482f.). And occasionally the subject may shift form one plane to another. Hektor rages at one moment like a god (Ares), at the next moment like a fire (15.605f). The alternate subjects can be explained by the tendency of similes to accumulate at moments of emotional intensity or heightened pathos. Far more common than the alternation of subject within a simile is the compositional technique of pairing or even stringing together similes.[11] In such cases the subject matter usually varies, as for example from hunting, to fire, to domestic animals, to birds (17.722ff.). Variation in fact is one of the primary

---

[11] Moulton, *Similes* 27–33, discusses successive similes, including the longest string at 2.455ff. prior to the Catalogue.

effects of a pair or string of similes. The multiplication of images allows our eye to wander beyond the action at hand and to embrace the world around it by focussing on a succession of events or landscapes in the simile. The sequence of events or scenes transpires in a desultory manner and often with the abruptness that characterizes dreams. But in spite of the apparent randomness of imagery, pairs of similes may be stitched together by shared vocabulary.

Let us examine two examples. In the opening passage of Achilles' fight by the River a fire–locust simile (21.12ff.) and a dolphin simile (22ff.) intensify the depiction of panic among the Trojans pursued by Achilles during his *aristeia*. Although the similes treat the theme of flight and cowardice common in *aristeiai*, the unusual subject matter argues for their being specifically adapted to the watery setting of the fight. The first simile compares the Trojans' helplessness in trying to escape from the raving man to swarms of locusts fleeing a blaze. The insects "cower" (πτώσσουσι) in the water of a nearby river. In addition to sketching their fear the simile emphasizes in the resumptive clause how the river was "filled" with horses and men. The idea of "filling" now recurs in the body of the dolphin simile which describes how when smaller fish "flee" the dolphin they "fill" up the corners of the harbor. The associated cowardice of the Trojans is then expressed in the resumptive clause of the simile by the same (φεύγω) verb used previously of the locusts. The use of πτώσσω both for locusts and men is remarkable. Literally the verb, meaning "to crouch in cowardice," is inappropriate for winged insects, but effectively it brings the human subject closer to the swarms of insects. The pattern of overlap can be set out as follows:

| Fire–Locust simile | Dolphin simile |
|---|---|
| φευγέμεναι (13: simile) | φεύγοντες (23: simile) |
| πτώσσουσι (14: simile) | πτῶσσον (26: resumptive ) |
| πλῆτο (16: resumptive) | πιμπλᾶσι (23: simile) |

A second example of interlocking imagery, this time involving an animal simile (boar 12.146ff.) and a tree simile (12.132ff.), occurs in the *Teichomachia* as Asios disobeys Poulydamas' orders to the Trojans to leave behind their horses and storm the wall in mass formation with Hektor. Asios alone attacks one of the gates of the wall in his chariot, but encounters the leaders of the Lapiths, Leonteus and Polypoites. From the Catalogue we know them as the pair who repelled the hairy–beast men (2.740–46). As their epithet αἰχμητής (12.128) indicates, the Lapiths are renowned spearmen. The second of the two similes honors this attribute by comparing the clatter of the tusks of two boars ripping up timber "at the roots" to the clatter of missiles deflected by their bronze armor. Like boars with menacing tusks, the Lapiths stand their ground and render impotent the enemy spears through their superior spearsmanship. Now the adjective πρυμνός is an epithet for δόρυ meaning "spear," or properly the wooden shaft of the spear. The subject of the earlier simile is a pair of oaks, δρύες, a cognate of δόρυ. In that simile we hear also of the roots of the tree, but the point is inverted: unlike the timber that is deracinated by the boars' tusks, the long roots of the oaks cling fast. This quality of steadfastness, of course, is the overriding point of comparison in both the oak and the boar simile. But the two similes are fundamentally intertwined with one another and linked to the narrative through the image of the spear, wherein resides the Lapiths' power of defense. The focus on the wooden spear and its analogues in the similes is reinforced by repeated vocabulary describing the general wooded setting: οὔρεσιν (132) – ὄρεσσιν (146). Having established the unity of the two similes we can observe a remarkable counterpoint of modifiers. The trees are depicted as "rearing their crests," ὑψικάρηνοι, like an animate creature[12], just as ἐκτάμνοντες, used of the cutting action of the boars' tusks, more appropriately belongs to the action of the woodcutter (δρυτόμος) in tree similes: "As the tumult goes up from men who are cutting | timber in mountain valleys, and the sound is heard far off..." (16.633f., cf. 11.86). In fact, the boar simile in Book 12 is an adaptation of the simile type describing the tremendous roar of trees falling in the forest. This second example shows that when one simile type which is a variation on another is juxtaposed with the original type, the simile pair and narrative achieve a heightened degree of unity.

## Simile and Omen

In Sanskrit literature similes occur in incantations, where the verbal device furnishes a parallel to a phenomenon in nature in order to influence the analogous subject whom a magician wants to affect.[13] In the Homeric simile the purpose of equating a warrior with a force in nature is never ex-

---

12 Cf. the periphrastic ἵππων ξανθὰ κάρηνα, 9.407.

13 Cf. J. Gonda, *Remarks on Similes in Sanskrit Literature* (1949) 74f.

plicitly anything but poetic, but implicitly a simile may confirm or predict events in the plot. The poet in a sense is like the bird interpreter, and his simile stands in the same relation to the narrative as an actual omen does to reality.[14]

Omens and similes have a similar verbal structure and function, as well as affinities in theme and content. In form omens resemble similes with an introductory and resumptive ὡς ... ὡς. In the *Iliad* there are three bird omens: 2.308ff., 8.247ff., 12.200ff.[15] The first two bear a resemblance to similes in form, theme, and linguistic detail, the last only in theme and in linguistic detail. In the first two the manifestation is described and its meaning then interpreted. The seer repeats the essential elements of the portent and relates them to the situation at hand by means of conjunctive ὡς. The simile form is especially marked in Kalchas' interpretation at 2. 326–29.

Omens and similes also bear a strong resemblance in content and theme to the mother–child relationship common in animal similes. 12. 200ff. describes an eagle carrying in his sharp talons a blood–red snake, but the snake bites the breast of the eagle, and the bird is forced to drop his victim. The snake writhes in the midst of the Trojan spectators. The emphasis of Poulydamas' interpretation lies in the eagle's failure to return home to feed the snake to her young (12.221f.):

> ... ἄφαρ δ' ἀφέηκε, πάρος φίλα οἰκί' ἰκέσθαι,
> οὐδὲ τέλεσσε φέρων δόμεναι τεκέεσσιν ἑοῖσιν

The omen is reminiscent of a bird simile (9.323f.) in Achilles' self–pitying speech, where the hero compares himself to a mother bird who brings back morsels for her young but is herself deprived of food. The passage at 2.308ff. tells of a blood–red snake attacking birds. Again mother and children are the subjects. The snake devours the eight νήπια τέκνα of a mother sparrow. The mother hovers above the empty nest and mourns the loss of her children. But the snake gobbles up the mother as well, making her the ninth victim. Again the theme and content of the omen resembles that of similes, such as the vulture simile at *Od.* 16.216ff., where two vul-

---

[14] On similes and omens in Homer and Hesiod, see S. H. Lonsdale, "Hesiod's Hawk and Nightingale: Fable or Omen? *Op.* 202–213," *Hermes* 117 (1989) 403–12.

[15] A. Podlecki, *G&R* n.s. 14 (1967) 12–23, shows that in the *Odyssey* both omens and dreams are closely connected with the bird similes (e.g., the vulture simile marking the reunion of Odysseus and Telemachos) and are interwoven into the imagery that embellishes the series of recognition scenes.

tures mourn the loss of their young (τέκνα) whom the farmers steal before they even had wings to fly; or the simile at 11.113ff. in which the lion devours the young (νήπια τέκνα) before the eyes of the helpless doe, who herself flees.

At 8.247ff. Zeus sends a sign of confirmation in response to Agamemnon's prayer. An eagle deposits a fawn at the altar of Zeus as the Achaians are sacrificing. The Achaians recognize that Zeus has sent the eagle and their confidence is restored: μνήσαντο δὲ χάρμης. As in the similes the fawn here is a symbol of helplessness and cowardice. The internal pictorial content of the portent is nearly replicated in the eagle simile at 22.308ff. in which the predatory bird swoops down upon a lamb or rabbit, which, like the fawn, are standard symbols of cowardice. Thus in language too omens are like similes, especially in the use of the emotive vocabulary so common in the animal similes. Both the omens in Books 8 and 12 contain vocabulary relating to heroic ideals: the snake has not forgotten his warcraft, and the portent makes the Achaians mindful of their courage.

Omens are related formally and thematically to similes and share significant vocabulary. To what extent can their functions be said to be similar? Do similes, like omens, actually predict or confirm narrative action? A group of three interrelated lion similes in the *Patrokleia* suggests that certain similes potentially behave like omens in battle narrative.[16] Hektor and Patroklos fight at close quarters twice, once over the corpse of Hektor's henchman, Kebriones, and once in the final duel. In the case of the former encounter, the first of a pair of lion similes (16.752f.) compares Patroklos about to seize Kebriones' armor to a wounded marauding lion "whose strength destroys him." In the resumptive clause the poet addresses Patroklos directly using the vocative. The second in this pair of similes (756ff.) casts Hektor and Patroklos as two lions fighting for the possession of a stag carcass. This second simile is organically interdependent with the final one of the Book, the famous lion–boar simile (823ff.) that spells Hektor's supremacy over Patroklos.[17] The earlier simile stating the lion's death can be seen as an example of foreshadowing; its effect is reinforced by the

---

[16] M. Baltes, *A&A* 29 (1983) 36–48, discusses the lion simile at 16.756ff. in relation to the earlier simile of two vultures (16.428ff.) and the lion–boar simile. He demonstrates that they form a sequence leading up to the death of Patroklos.

[17] Cf. the repetition of the dual in ὥ τ' ὄρεος κορυφῇσι at the beginning of 757 and 824, and μέγα φρονέοντε μάχεσθον at the end of 758, 824. In the first simile two lions fight over a carcass; in the second a lion and boar contend over a mountain spring.

lion–boar simile, the outcome of which signals a reversal. Hektor as victor assumes the image of the unvanquished lion. All warriors implicated in the plan of Zeus, in fact, are allotted similes in which the death or potential death of the animal is mentioned.

Other similes confirm action in the narrative. There are two similes in which the objects of comparison are said to lose their courage after being compared disadvantageously to the victims of vicious animals. The phrasing of the first, a bird simile, resembles omens we have already looked at. At 17.755ff., when the Achaians see Hektor charging with Aineias, they lose their courage and cry aloud; and their crying is like the screams of smaller birds when they see the hawk, the bird of Apollo, and Hektor's patron divinity, "who brings death."

> τῶν δ' ὥς τε ψαρῶν νέφος ἔρχεται ἠὲ κολοιῶν      17.755
> οὖλον κεκλήγοντες, ὅτε προίδωσιν ἰόντα
> κίρκον, ὅ τε σμικρῇισι φόνον φέρει ὀρνίθεσσιν,
> ὣς ἄρ' ὑπ' Αἰνείαι τε καὶ Ἕκτορι κοῦροι Ἀχαιῶν
> οὖλον κεκλήγοντες ἴσαν, λήθοντο δὲ χάρμης.

In the Sarpedon–lion simile at 12.299ff. Sarpedon holds his shield out in front of him about to attack. A short, emblematic lion simile precedes the description of this action. A longer lion simile follows. The θυμὸς ἀγήνωρ of the lion bids him to attack the sheepfold, just as the θυμός of Sarpedon urges him to leap over the wall: θυμὸς ἀνῆκε (307). The spirit of the warrior fuses with the spirit of the lion.

In these examples the poet's pronouncement of the simile closely aligns the event in nature with the human action. The poet is similar to the bird prophet who has the insight to perceive the manifestation of the human in the natural world and the ability to express the parallel in metrical verse. He employs a simile or omen to elucidate or shape narrative events, much as the prophet uses an omen to express or control objective occurrences in reality.

## Simile and Epiphany

The related question of bird epiphanies is of interest, not only because of their possible implications for the history of religion,[18] but because of

---

[18] M. P. Nilsson, *Geschichte der griechischen Religion*[3], *Hdbch. d. Altertumswiss.* 5 II (1967) 290–92; E. T. Vermeule, *Götterkult, Archaeologica Homerica* V (1974) 94.

what they can tell us about beliefs underlying the similes.[19] Recently Hartmut Erbse has reexamined the evidence for transformations of divinities into human as well as bird and other animal forms. He has conclusively shown that the poet has degraded actual theriomorphosis of divinities into a disguise with which they conceal their identities.[20] Six passages in the *Iliad* and *Odyssey* likening gods to birds (7.59, 14.290, *Od.* 1.319, *Od.* 3. 371f., *Od.* 5.353, and *Od.* 22.239[21]) may indicate either an actual transformation or merely a comparison between bird and divinity. The terms used in the passages, εἰδόμενος, ἐοικώς, and ἐναλίγκιος, all admit an ambiguity. The transmutation of gods into human and other shapes argues in favor of actual avian metamorphosis by gods in Homer. When gods transform themselves into human shapes, they do so in such a way as to deceive their mortal onlookers. Then the poet uses the participle εἰδόμενος (e.g., *Od.* 2.267f., for Athene–Mentes), which is also used when Athene departs in the form of the φήνη (sea–eagle[?], *Od.* 3.372). As an example of a divinity taking animal shape, we have the sea–god Proteus. It seems best to take at face value the description of Apollo and Athene as birds, perched high in Zeus' oak to observe the fighting (7.58–61),[22] as well as

---

See W. Burkert, *Greek Religion*, tr. J. Raffan (1985) 64–66, on the relationship between theriomorphism and animal worship. See also J. Pollard, *Birds in Greek Life and Myth* (1977) 155–159.

[19] F. D. Dirlmeier, *Die Vogelgestalt homerischer Götter*, SHAW (1967) 2, conveniently reviews the literature surrounding this controversy since the time of Heyne, and in agreement with the latter's remarks on 7.58f. (*Homeri Carmina V* [1802] *ad loc.*) argues that all examples of alleged bird epiphanies are to be taken no differently from any other type of short simile. The force of the simile, he asserts, lies in a comparison, implicit or explicit, between the bird's speed and the god's swift movement. He cautions strongly against interpreting the similes as reminiscences of Minoan bird epiphanies or as evidence of theriomorphism; cf. Fränkel, *Gleichnisse* 81f. and Schnapp–Gourbeillon, *Lions* 185–90.

[20] "Der Dichter hat die Tiergestalt, Residuum einer älteren, aber im archaischen Griechenland noch immer lebendigen Form der Gottesvorstellung, zur blossen Maske gewissermassen degradiert, deren Wahl dem Ermessen des jeweiligen Gottes unterliegt." H. Erbse, "Homerische Götter in Vogelgestalt," *Hermes* 108 (1980) 273. Erbse's theory that bird epiphanies are the residue of theriomorphosis accords with the interpretation (e.g., Kirk, *Commentary* 110) that animal–related epithets for divinities (e.g. βοῶπις for Herē, γλαυκῶπις for Athene) are vestiges of Mycenaean–Minoan religious conceptions or artistic representations of the divinities concerned in animal forms.

[21] The simile at 13.62ff. comparing Poseidon to a hawk as he descends to cheer the Aiantes permits no ambiguity, since Aias, son of Oileus, says he saw the feet and legs of the departing figure. The translation "talons and tarsus" for "feet and legs," suggested by J. Pollard, *Birds in Greek Life and Myth* 158, strains the Greek.

[22] Cf. H. Bannert, "Zur Vogelgestalt der Götter bei Homer," *WS* 91 (1978) 42, who rejects Dirlmeier's rationalizing interpretation.

that of Sleep (who is portrayed as winged in Classical art), alighting on the branches of a pine in the likeness of the chalkis–kymindis (14.290). Two problematic similes are found at *Od.* 1.319–323 and *Od.* 3.372f., where Telemachos and the suitors, and later Nestor, express surprise at the departure of Athene, compared to a bird. Are they surprised because they see the human form (which has moments before been speaking) suddenly vanish, leaving them half–aware that a god may have been in their midst? Or are we to believe that the spectators suddenly find themselves looking at a bird taking wing? Herē's remark to Poseidon during the encounter between Aineias and Achilles is telling, "it is hard for gods to be known in their true shape" (20.131). As in the opening of *Odyssey* 14, where only Odysseus and the dogs are aware of Athene's presence, the sighting of a god in any form is probably left only to those spectators intended by the divinity and allowed for by the dramatic requirements of the plot. It will never be possible to know the exact nature of beliefs about metamorphosis among the epic audience. The Proteus episode, however, provides a unique opportunity to compare the various types of theriomorphosis found in the early Greek epic in the same passage: the supernatural (Proteus), the literal (disguise in seal skins), and the metaphorical (the lion simile). Analysis of the interaction of these three types indicates that all share a basis in beliefs in metamorphosis.[23] The proportionately greater quantity of metamorphoses in the Epic Cycle makes it tempting to ask if the frequent animal similes in Homer provide the vehicle for re–establishing a lost "magical" dimension.[24]

## Simile and Ecphrasis

Like omens and similes, ecphrastic descriptions of artistic works may also function as signs. Above we observed the interplay between Φόβος on Agamemnon's telamon and in a lion simile in his *aristeia*.[25] The scene described on Odysseus' fibula is another example of the kinship of simile

---

[23] Cf. S. H. Lonsdale, "Protean Forms and Disguise in *Odyssey* 4," *Lexis* 2 (1988) 165–78.

[24] See J. Griffin, "The Epic Cycle and the Uniqueness of Homer," *JHS* 97 (1977) 39–53; cf. R. Janko, "The Shield of Heracles and the Legend of Cycnus," *CQ* (1986) 52–53.

[25] Pp. 57–58.

and ecphrasis.[26] The fibula is one which Odysseus wore to fasten his brilliantly colored cloak of wool on that day twenty years ago when last entertained by his host, who is none other than Odysseus himself, now disguised as a beggar. Speaking to Penelope, he describes in coy detail the skilful craftsmanship of the ornament and the wondrous work of art fashioned upon it—a dog grasping a struggling fawn in its forepaws and holding it with its gaze; all those who admired it marveled at how, though made of gold, the hound held fast the fawn as it struggled to escape. Odysseus' evocation of the gold heirloom has the intended effect of further piquing Penelope's curiosity, and it no doubt delighted the audience, who appreciated such descriptions of the natural world for their own sake. This vignette of the hunt resembles scenes of pursuit in the *Iliad* similes analyzed above. For example, at 15.579ff. Antilochus leaps to seize the armor of his opponent, just as a hound springs to attack a fawn wounded by a hunter's arrow. Odysseus and Diomedes, tracking Dolon, are compared to two hounds chasing a fawn or hare (10.360ff.). There is no essential difference in content, language, or function between the hunt described on the fibula or in the similes. Both are pictorial fragments cut loose from the hunt. As with the similes portending a warrior's fate in the *Iliad*, the import of the dog's mastery of the fawn foreshadows Odysseus' eventual victory over the suitors.[27] The familial likeness of lion similes to vignettes from the Shield, the most extended example of ecphrasis in either poem, is a further example of the overlap of simile and ecphrasis discussed in the next section.

### Lion–Cattle Similes and the Theme of the Cattle Raid

As the interaction between the metaphorical and literal uses of "shepherd" indicates, herding similes link the pastoral theme to the military subject of the epic.[28] The epic theme of the cattle-raid is also adapted to the anticipated plunder of Troy through the lion–cattle similes where cattle represent

---

[26] J. L. Myres, "Homeric Art," *ABSA* 45 (1950) 229–260, compares the fibula with seal–devices; cf. E. T. Vermeule, *The Art of the Shaft Graves of Mycenae* (1975) 41f., n.; R. A. Prier, "That Gaze of the Hound; *Odyssey* 19.228–231," *RhM* 123 (1980) 178–180.

[27] For other signs of Odysseus' successful rout of the suitors, see G. P. Rose, *TAPA* 109 (1979) 215–230.

[28] Cf. Moulton, *Similes* 117–19, who shows that many of the *Odyssey* similes are related to major epic themes.

material possessions and social values.[29] In the epic a simple formula, akin in form and content to measurement similes, exists for reckoning a man's holdings. It begins by stating the number of herds of cows and then repeats the quantitative adjective τόσ(σ)α before the various remaining flocks and herds. Thus Eumaios tallies up Odysseus' livestock holdings on the mainland by stating that his master owns twelve herds of cattle, as many flocks of sheep and as many droves of swine, and as many herds of goats (*Od.* 14.100f.). The formula is well–established in Homer. Even Helios' herds and flocks are reckoned in this way (*Od.* 12.128-30), as well as the impressive numbers of animals that Nestor boasts of exacting in reparation from the Eleians, who had stolen cattle from Neleian Pylos (11.679-81).

The importance of cattle in Homeric society is made clear throughout the *Odyssey*, where flocks are a barometer of social harmony and rectitude. The well–known king simile in Odysseus' speech to Penelope at *Od.* 19.109ff. foreshadows the order that Odysseus will restore to Ithaka once he has reinstated himself as lawful ruler. Among the signs of prosperity mentioned is continuous birthing among the sheepflocks. Disrespect for another man's livestock signals social disorder. Virtual anarchy reigns in Ithaka in Odysseus' absence. The divergence from the social norm is underscored by the suitors' devastation of another man's property. The

---

[29] Cattle are also important sacrificial animals, both in a funerary context (e.g. the burial of the dead 7.466; funeral of Patroklos 23.30), and as a propitiatory offering to the gods, as Nestor so elaborately demonstrates to Telemachos at *Od.* 3.436ff. Even on the battlefield the cow's presence is felt. The many layers of cowhide pad out the round shield. Hektor knows how to "plough" his "dry cow" through the ranks of men, νωμῆσαι βῶν | ἀζαλέην (7.238f.). In the absence of circulating currency cattle were a measure of value and played an important role in the social institution of marriage in wooing gifts. The practice is alluded to in a biographical digression in the *Iliad* and in several narrative passages in the *Odyssey*. In the *Iliad* the suitor offers cattle to the bride's family (11.242–45, cf. the sarcastic ἐέδνωταί, 13.382). In the *Odyssey* (ἔ)εδνα refers to both the wooing gifts which the suitors compete among each other to present for Penelope's hand (e.g. *Od.* 15.17ff.) and to the dowry prepared by the bride's family (*Od.* 1.277, 2.19). See Stanford, Commentary *ad loc.*, who believes that there is an ambiguity in *Od.* 1.277. He argues that οἱ δέ could refer to the suitors rather than to the bride's family, but this puts an unnecessary strain on the Greek. He does not mention the divergence between the *Iliad* passage and the *Odyssey* passage. For a more detailed discussion, see M. I. Finley, "Marriage, Sale and Gift in the Homeric World," *Rev. Internat. des Droits de l'Antiquité*, 3rd ser. 2 (1955) 167–94; cf. W. K. Lacey, "Homeric ἔδνα and Penelope's κύριος," *JHS* 86 (1966) 55–68. A. M. Snodgrass, *JHS* 94 (1974) 114–125, in discussing marriage customs (115ff.), calls in anthropological evidence to show that division into bride–price and dowry practices is misleading. I. Morris, *CA* 5 (1986) 105–15, reassesses the scholarly debate on dowry. For the scholiasts' view of ἔδνα, see M. Schmidt, *Die Erklärungen zum Weltbild Homers und zur Kultur der Heroenzeit in den bT – Scholien zur Ilias, Zetemata* 62 (1976) 240–247.

breach of custom which the suitors' lawless consumption of Odysseus' cattle signifies in the *Odyssey* finds its parallel in the *Iliad* in the illegal possession of cattle. Achilles implies at 1.154ff. that the theft of another man's livestock constitutes grounds for retaliation. But even though cattle–reiving was frowned upon, one gains the impression that rustling was quietly condoned and even encouraged, since it may have been an initiation rite, as suggested by Nestor's recital in Book 11.[30] In revenge for King Augeias' failure to return the horses and chariot which Neleus had sent to compete in the games at Elis, Neleus raised a force of men and managed to abscond with fifty herds of oxen, as many sheep flocks and droves of pigs, and again as many herds of goats, and 150 horses. A counter–attack is launched by the Epeians, and fighting breaks out in Pylos. The length of digressions like Nestor's and the frequent flashbacks to Achilles' extra–Trojan deeds as a rustler in Lyrnessos or Pedasos indicate that cattle–raids were a subject of abiding interest to Homer's audience. Like the bride–raid, the cattle–raid was a major epic theme.

The theft of cattle, for which Homeric Greek had the compound βοη‑λασίη, was the stuff epic poetry was made of in ancient Greece and other heroic cultures, notably Celtic and Indian.[31] Hesiod (*Op.* 161–65) attributed the perdition of the race of heroes to two underlying causes: the Trojan war precipitated by the theft of Helen and the fight over the cattle of Oidipus in Thebes.[32] Elsewhere in early Greek hexameter poetry the cattle raid is a subject in the Hesiodic *Catalogue of Women* (fr. 256 Merkelbach–West), the pseudo–Hesiodic *Aspis*, and the *Homeric Hymn to Hermes*.[33] As we have seen, the plot of the *Odyssey* is set against the backdrop of the gradual theft of Odysseus' livestock by the suitors and the eventual battle

---

[30] P. Walcot, "Cattle Raiding, Heroic Tradition, and Ritual: The Greek Evidence," *History of Religions* 18 (1979) 326–351, speculates that the recurring detail of the solitary survivor (in Nestor's story 11.693; cf. *Aspis* 14ff., Apollodorus 2.4.6) may be an indication of a pattern or folkloric motif in accounts of cattle–reiving. Kirk, *Songs* 327, states, "the Nestor reminiscences presuppose a special source, a Pylian song or group of songs." For an interpretation of the function of this episode and a review of the scholarship, see V. Pedrick, *TAPA* 113 (1983) 55–68.

[31] See J. Weisweiler, "Vorindogermanische Schichten der irischen Heldensage," *Zeit. f. Celtische Philologie* 23 (1954) 27–28; A. Venkantasubbiah, "On Indra's Winning of Cows and Water," *Zeit. d. deutschen morgendländischen Gesellschaft* 115 (1965) 120–33.

[32] By the latter Hesiod presumably meant the struggle for kingship between Eteocles and Polyneices that broke out over Oidipus' wealth, measured in terms of cattle.

[33] See S. Shelmerdine, *TAPA* 116 (1986) 57f., on the resemblances between the theft of Apollo's cattle and the theft of Helios' cattle.

that erupted when Odysseus punished them for the depredation of his flocks.[34]

In the *Iliad* the theme of the cattle–raid occurs not only in digressions but in the condensed form of twenty–two lion similes in which the lion preys on cattle.[35] Two scenes from the Shield of Achilles demonstrate the proximity of human and animal predation in two parallel scenes, both centering around cattle feeding and watering at a river. The first (18.520–40), from the "City at War," describes an ambush in which a group of reivers cut off the cattle on both sides and kill the shepherds. The second, from the agricultural vignettes (18.573–86), rephrases the first scene in the animal metaphor. Instead of reivers two marauding lions steal upon a herd of oxen and capture a bull while the farmers and dogs protest in vain.

The resemblances between the two vignettes are detailed and specific, and upon closer inspection they form a kind of doublet which shares an affinity with the lion–cattle similes through common language. Both attack scenes take place near a river (521, 576); the rushing sound of the river (κελάδοντα, 576) echoes the loud uproar (πολὺν κέλαδον, 530) that alerts the shepherds to the ambush. In both vignettes the number of shepherds (called νομῆες) is given (δύω, 525, τέσσαρες, 578). The poet takes care to represent equally the concerns of both sides. We hear, of course, about the stealth and aggression of the attackers, but we also learn of the shepherds' reaction. In the first vignette they are caught off guard because they were enjoying their reed pipes and did not expect any deceit, δόλον δ' οὔ τι προνόησαν (526). Finally they are killed. The inclusion of such details told from the point of view of the helpless defender to introduce a note of pathos recalls the lion similes, where the shepherd is said, for example, to be absent (15.325), or only successful in grazing the intruder while he watches his herds being ravaged (5.136ff.). In the second vignette we hear of the brave but futile attempts of the shepherds to set their hounds on the lions (583ff.). From the perspective of the attacker, we hear the graphic details of their dirty work as the ambushers surround and cut off the cattle, then kill the shepherds; or of the lion seizing a bull, then ripping open the hide of a large cow to devour its blood and entrails. The phrase describing their bloody feast closely resembles the phrase αἷμα καὶ ἔγκατα πάντα λαφύσσει, which is otherwise restricted to but formulaic

---

[34] The *Odyssey* (12.353-65) also mentions the theft of Helios' cattle.

[35] 5.136, 161, 556, 11.172, 383, 548, 12.293, 299, 15.324, 586, 630, 16. 487, 17.61, 109, 542, 657, 18.161, 20.164, 22.262, 24.41, *Od.* 6.130, *Od.* 22. 402.

in lion similes.[36] The cattle raid and attack on cattle by the predator are virtually interchangeable in the epic tradition. Priam applies the term ἁρπακτῆρες (24.262) to his good–for–nothing children, who "would even steal lambs and kids from their own people." The word is derived from the verb ἁρπάζω, used of the lion seizing its victims.[37]

Once a digression on cattle raiding is juxtaposed with a marauding lion simile. Before the confrontation between Achilles and Aineias, the latter recalls to Apollo (in the guise of Lykaon) that Achilles "came after our cattle | the time he sacked Lyrnessos and Pedasos."[38] Prior to addressing Aineias, Achilles is compared to a marauding lion in an elaborately developed simile. Like the Kalydonian boar that had posed a threat to the stability of the region, the whole country gathers up in arms against the beast (20.165–179). The occurrence of the simile in the vicinity of the cattle raid suggests that when he returns to the fighting, Achilles the cattle–rustler metaphorically becomes the marauding lion.

As observed in Chapter 4, the parasitic depredation of the flocks of men is a metaphor for the Trojan conflict in general. The Achaians are the aggressors who have strayed from their homeland. Their forces lie encamped on a thin strip of land at the periphery of the citadel; they are eager to seize the possessions within. The Trojan defenders are comparable to the farmers defending their flocks. The livestock correspond to the women and children and all that is valuable within the citadel. The enceinte wall surrounding the citadel is represented in the similes by the enclosure about the cattle–steading. The lion–cattle similes cumulatively form a motif that metaphorically restates the Achaians' desire to "leap over" the walls of the Trojan citadel and capture the possessions within. The motif, however, is versatile. When the dramatic situation shifts from an Achaian attack on the citadel to the Trojans' attack on the Achaian wall and ships in Book 12 and 15, lion imagery temporarily gravitates toward the Trojans.


## Simile and Paradigm

The resemblance between the cattle–raid and the lion–cattle similes suggests that some similes may, like Nestor's digression in Book 11, have

---

[36] 11.176=17.64; πρώτῃσι βόεσσι(ν) 18.579, cf. 15.634.

[37] 5.556, 12.305, 13.199, 18.319.

[38] 20.91ff. Aineias, who was caught apart from his cattle by Achilles, is accorded the epithet "shepherd of the people" at 20.110.

a paradigmatic function. We have seen that often a lion simile is meant to exhort a warrior to attack like the beast. We can observe a similar congruence between the Meleager paradigm and the boar similes. As Willcock and others have shown, paradigms in Homer are strongly characterized by ring composition, a form we have observed in similes as well.[39] If we take the events leading up to the dispute over the spoils from the boar (9.529–49) and compare it with boar similes, similarities in form and content emerge.

ἡρώων, ὅτε κέν τιν' ἐπιζάφελος χόλος ἵκοι·     9.525
δωρητοί τε πέλοντο παράρρητοί τ' ἐπέεσσι.

...

Κουρῆτές τ' ἐμάχοντο καὶ Αἰτωλοὶ μενεχάρμαι
ἀμφὶ πόλιν Καλυδῶνα καὶ ἀλλήλους ἐνάριζον,     530
Αἰτωλοὶ μὲν ἀμυνόμενοι Καλυδῶνος ἐραννῆς,
Κουρῆτες δὲ διαπραθέειν μεμαῶτες Ἄρηϊ.
καὶ γὰρ τοῖσι κακὸν χρυσόθρονος Ἄρτεμις ὦρσε
χωσαμένη ὅ οἱ οὔ τι θαλύσια γουνῶι ἀλωῆς
Οἰνεὺς ῥέξ'· ἄλλοι δὲ θεοὶ δαίνυνθ' ἑκατόμβας,     535
οἴηι δ' οὐκ ἔρρεξε Διὸς κούρηι μεγάλοιο.
ἢ λάθετ' ἢ οὐκ ἐνόησεν· ἀάσατο δὲ μέγα θυμῶι.
ἣ δὲ χολωσαμένη δῖον γένος ἰοχέαιρα
ὦρσεν ἔπι χλούνην σῦν ἄγριον ἀργιόδοντα,
ὃς κακὰ πόλλ' ἔρδεσκεν ἔθων Οἰνῆος ἀλωήν·     540
πολλὰ δ' ὅ γε προθέλυμνα χαμαὶ βάλε δένδρεα μακρὰ
αὐτῆισιν ῥίζηισι καὶ αὐτοῖς ἄνθεσι μήλων.
τὸν δ' υἱὸς Οἰνῆος ἀπέκτεινεν Μελέαγρος
πολλέων ἐκ πολίων θηρήτορας ἄνδρας ἀγείρας
καὶ κύνας· ...     9.545

...

ἀλλ' ὅτε δὴ Μελέαγρον ἔδυ χόλος, ὅς τε καὶ ἄλλων
οἰδάνει ἐν στήθεσσι νόον πύκα περ φρονεόντων,
ἤτοι ὃ μητρὶ φίληι Ἀλθαίηι χωόμενος κῆρ     555
κεῖτο παρὰ μνηστῆι ἀλόχωι καλῆι Κλεοπάτρηι

cf.

ὡς δ' ὅτε κάπριον ἀμφὶ κύνες θαλεροί τ' αἰζηοὶ     11.414
σεύωνται, ὃ δέ τ' εἶσι βαθείης ἐκ ξυλόχοιο
θήγων λευκὸν ὀδόντα μετὰ γναμπτῆισι γένυσσιν,
ἀμφὶ δέ τ' ἀΐσσονται, ὑπαὶ δέ τε κόμπος ὀδόντων

---

[39] M. M. Willcock, *CQ* 14 (1964) 147.

> γίγνεται, οἳ δὲ μένουσιν ἄφαρ δεινόν περ ἐόντα,
> ὥς ρα τότ' ἀμφ' Ὀδυσῆα Διὶ φίλον ἐσσεύοντο
> Τρῶες·...

cf.

> ... ἀτὰρ Δαναῶν γένετο ἰαχή τε φόβος τε,    12.144
> ἐκ δὲ τὼ ἀίξαντε πυλάων πρόσθε ,μαχέσθην
> ἀγροτέροισι σύεσσιν ἐοικότε, τώ τ' ἐν ὄρεσσιν
> ἀνδρῶν ἠδὲ κυνῶν δέχαται κολοσυρτὸν ἰόντα,
> δοχμώ τ' ἀίσσοντε περὶ σφίσιν ἄγνυτον ὕλην
> πρυμνὴν ἐκτάμνοντες, ὑπαὶ δέ τε κόμπος ὀδόντων
> γίγνεται εἰς ὅ κέ τίς τε βαλὼν ἐκ θυμὸν ἕληται·    12.150
> ὣς τῶν κόμπει χαλκὸς ἐπὶ στήθεσσι φαεινὸς
> ἄντην βαλλομένων· ...

The theme of anger is mentioned prior to and following this passage at 523, 525, and 553, 555; the mortal anger of Meleager and earlier heroes and Althaia (χόλος–χωόμενος, 525, 555) frames the divine anger (χωσαμένη–χολωσαμένη, 534, 538) which causes the goddess to rouse the boar to destroy Oineus' fields. ὦρσεν for the dispatching of the boar, and κακά for the damage it does (539f.), have been anticipated by κακὸν ... Ἄρτεμις ὦρσε at 533, which describes the outcome in a compressed manner. Another ring is formed by the confrontation between Meleager and the boar and the subsequent quarrel between the Aitolians and Kouretes over the head and hide of the boar. The confrontation between Meleager and the boar is expanded by a description of the boar tearing up the trees and flowers at the roots and Meleager's efforts to gather up a troop of men and dogs to repulse the beast. The same basic elements—men and dogs confronting a boar uprooting the underbrush—form the internal pictorial content of three boar similes (11.414ff., 12.146ff., 13.471). The first of these is framed by the repeated action verb σεύω (11.415, 419). The second has an even more pronounced ring, like that in the paradigm, based on a comparison between the battle of the warriors in the narrative (ἰαχή, 12.144, cf. κόμπει χαλκός 151) and the noise and confusion caused by the boar's uprooting activity (κόμπος ὀδόντων, 149). We can observe that the belligerent action of the animal in the Kalydonian boar hunt becomes the central scene in a simile type. If, as Willcock and others have argued,[40] the compressed nature of this portion of the Meleager paradigm means that the poet is abbreviating an already existing poem, we have in

---

[40] Willcock, *op. cit.* 149, n.1.

the boar similes an even more telescoped version. What we are left with is not a description of the anger of Meleager or Artemis, but that of the boar as it angrily uproots the undergrowth. The vivid depiction of the boar's furor in a simile is then metaphorically transferred back to the human counterpart in the narrative in such a way as to exhort him to unleash his anger against his opponent.


## Conclusion

An examination of the position of animals in Homer suggests the fundamental and sometimes intimate relationship between man and animal. Economically animals represent the measuring–stick of material wealth. In sacrifice, as well as in bird divination, the animal holds a mysterious power for man as he tries to communicate with the gods. Hunted animals, particularly the boar, are worthy opponents for heroes. In the pastoral setting a deep intimacy is shared by herdsman and herd, and the farmer relies on his watchful dog. The portrayal of the animal in Homer is consistent and straightforward, with attention paid to naturalistic portrayal. Where animals occur in mythological contexts the supernatural element is subordinate. However, the close resemblance between similes and omens and bird epiphanies suggests that similes may at an earlier stage have had a more overtly supernatural meaning. An abundance of epithets for animals indicates description for description's sake, and the outstanding features that capture the poet's eye is animal motion with which he choreographs the movements of warriors and groups of men in similes. A unique feature of animal similes is the use of heroic and emotive vocabulary for purposes of intensifying pathos. The anthropomorphic portrayal of the animal results in linguistic unity between simile and narrative. The overlap of animal imagery and narrative at the level of diction is complemented by resemblances between simile and other contexts in form and theme. The various components of the animal imagery work together to orchestrate a subliminal concert that resonates with the poem.

The themes which recur throughout the animal similes express human concerns: the protection of young, separation, defence, the triumph of the strong over the weak. Wasps rush out of their hives to protect their children, Odysseus' heart growls like a bitch protecting her puppies when a stranger intrudes,[41] vultures cry for their young whom farmers have stolen from the nest. Similes of pursuit, especially the marauding lion similes and hunting similes, recast the aggressive urge in various forms. These themes, which provide the fundamental link between the animal and human worlds, pervade the epic: Hektor leads the defence of the Trojans in order to protect the women and children within the citadel; Andromache and Hekabe mourn Hektor; in Book 11 Odysseus, even though he is eventually wounded, exemplifies the ideal of self–defence in the face of attackers when cut off from his companions, and Aias defends his brethren in the ensuing episode by sending the Trojans into panic; the wrath of Achilles attains monumental proportions only after he has suffered loss and separation from Patroklos; he in turn hunts down the killer. In the *Odyssey* the majority of the similes revolve around the themes of separation and recognition. The simile comparing Telemachos' and Odysseus' cries of joy upon their reunion to the shrill lamentation of the vultures bereft of their young (*Od.* 16.215ff.) is a case in point. A parallel example occurs at *Od.* 10.408ff. After the companions of Odysseus are freed from the magical spell of Circe, they see their familiar leader and are compared to lowing calves huddling around their mothers as they return from pasture. The welcome sight of Odysseus evokes a strong response in the heroes, who cry and imagine themselves at home in more habitual surroundings.

The simplicity and universality of these themes in both human and animal communities make them ripe for poetic treatment in an epic where the role of comparison between both spheres is established through the stylistic device of the developed simile. By means of an illuminating image, the simile momentarily suppresses the narrative event, transposing the action at hand to another register. The actor and the event are recast and reintegrated.

The epic tradition not only described the physical aspects of its characters but explored the spiritual dimensions as well. In so doing a singer often resorted to an animal simile to supplement portrayals of internal and spiritual states; the simile was also a cogent way of making the audience confront the universal suffering evoked by the death of even a minor warrior. The animal simile was a part of a poet's repertoire for explaining the meaning of death. As in the war between the Kouretes and the Aitolians

---

[41] Plutarch quotes this (*Od.* 14.14f.) and two other similes (9.323f., 17.133ff.) as examples of the self–sacrificing animal parent in the prologue to *De amore prolis* (494 C–E).

over the divinely dispatched boar, or as in Herakles' sack of Troy, a conflict or misunderstanding over animals is given as the cause for an ensuing war. The *aition* of war often involves the capture of animals, be it in cattle-raiding, horse–thieving, or hunting. Phoinix's elaborate introduction to the Meleager paradigm shows that stories explaining the origins of war were an epic theme (9. 524–28):

οὕτω καὶ τῶν πρόσθεν ἐπευθόμεθα κλέα ἀνδρῶν
ἡρώων, ὅτε κέν τιν' ἐπιζάφελος χόλος ἵκοι·
δωρητοί τε πέλοντο παράρρητοί τ' ἐπέεσσι
μέμνημαι τόδε ἔργον ἐγὼ πάλαι, οὔ τι νέον γε,
ὡς ἦν· ἐν δ'ὑμῖν ἐρέω, πάντεσσι φίλοισι.

The Kalydonian boar hunt is recalled by an episode in the seventh Book of the *Aeneid*. When war is about to break out between the Trojans and the Rutulians, Allecto inspires frenzy into the hounds and they run down a stag dear to Silvia. The young Ascanius fires the first fatal arrow and kills the stag. As the poet states, this is the first cause of turmoil that rouses the people of the countryside to war, "ut cervum ardentes agerent: quae prima laborum I causa fuit, belloque animos accendit agrestis" (*Aen.* 7.481f.). This remark reveals that a concrete cause for war is needed, and often that *aition* is centered around or transferred to an animal. Hunting and herding similes in the *Iliad* reproduce on a much smaller scale an explanation for the fighting in the narrative by providing an analogue from the lives of the common farmer or hunter.

The reduction of the magical element in Homer results overall in a pronounced distinction between man and animal in the narrative. In Circe's transformation of the companions of Odysseus into pigs, the point is made that in spite of their animal form the minds and thoughts of men remain the same (*Od.* 10.240). In the similes, the converse occurs: emotive vocabulary impresses upon the listener the similarity of internal states between beast and man. What emerges when a human psychological or emotional state is grafted onto the animal is a hybrid with no less poetic and dramatic force than the chimaera, Kerberos, centaurs, or satyrs. The hybrid in Greek myth and art had a special appeal and helped to explain and account for the unknowable. It gripped the imagination and provoked thought, just as the animal simile in its protean shapes invites many different literary and philosophical interpretations.

Animals in Homer served a quasi–historical function by explaining and illustrating the past. The animal stimulated a longing for the heroic age. Because the pastoral scene was basically static, it provided a tangible link with the heroic age which audiences imagined as preceding their own. Through the use of animal imagery there was a blurring of man, god, and animal that merged the audience with the heroic subject. This was accomplished in part by episodes accounting for the animal origins of strife, and in part by the transference of human emotions and situations to the animal world. In folklore animals appear in origin myths for death, dance, war, and the like to explain the beginnings of cultural and biological forms that are difficult to explain. By appealing to the familiar and commonplace, the poet transformed the animal world into a poetic language through which he explored the origins and nature of such fundamental human concerns as love, glory, war, and death.

# Appendix A
## A Register of Identifiable
## Species in the Homeric Poems

**Domesticated Animals**

| | |
|---|---|
| αἴξ (usu. f.) | goat |
|   τράγος | he–goat |
|   ἔριφος | kid |
|   (cf. μῆλα) | |
| βοῦς | cow |
|   (βοῦς) ταῦρος | bull |
|   βοῦς ἄρσην | steer |
|   πόρτις (f.) | calf |
|   πόρις (f.) | calf |
|   πόρταξ (f.) | calf |
| ἡμίονος | mule |
| ἵππος | horse |
|   ἄρσην | |
|   θήλεια | |
|   πῶλος (m., f.) | foal |
| κύων (m., f.) | dog |
|   σκύλαξ | puppy |
| μῆλον | sheep or goat; small cattle |
|   ἔμβρυον | young |
|   πρόγονος | early–born |
| ὄις (m., f.) | sheep |
|   κτίλος | lead–ram |
|   κριός | ram |
|   (ὄις) ἀρνειός | wether |
|   ὄις ἄρσην | wether |
|   ἀρνός | lamb, wether |
|   (cf. μῆλα) | |
| ὄνος (m., f.) | ass |

| | |
|---|---|
| οὐρεύς | mule |
| πρόβατα (n. pl.) | cattle |
| σ(ῦς) | pig |
|   χοῖρος | porker |
|   (σῦς) σίαλος | fat hog |

**Birds**

| | |
|---|---|
| ἀηδών | nightingale |
| αἰγυπιός | lammergeyer? |
| αἰετός | eagle |
| αἴθυια | shearwater (*lat. mergus*) |
| ἄρπη | unknown raptor |
| γέρανος (f.) | crane |
| γλαύξ (f.) | owl |
| γύψ | vulture |
| ἐρωδιός | heron |
| ἴρηξ | hawk, falcon |
| κήξ (f.) | tern |
| κίχλη | thrush |
| κίρκος | hawk, falcon |
| κολοιός | jackdaw |
| κορώνη | crow |
| κύκνος | swan |
| λάρος | gull |
| οἰωνός | bird |
| ὄρνεον | bird |
| ὄρνις | bird |
| πέλεια | dove |
| πετεηνά (n. pl.) | birds |

| | | | |
|---|---|---|---|
| σκώψ | scops owl | εὐλή | larva of wood–boring insect |
| στρουθός | sparrow | | |
| τρήρων (f.) | wild dove | ἴψ | wood–worm; moth-larvae[4] |
| φάσσα | wood–pigeon or ring–dove | | |
| | | κυνοραιστής | dog–flea |
| φήνη | ?lammergeyer | μέλισσα | bee |
| χαλκὶς κύμινδις | a variety of owl | μυῖα | fly |
| χελιδών | swallow | κυνάμυια | dog–fly |
| χήν (f.) | (wild) goose | οἶστρος | gad–fly[5] |
| ψάρ | starling | σκώληξ | earthworm |
| | | σφήξ | wasp |
| | | τέττιξ | cicada |

**Fish, Sea–beasts, etc.**

| | |
|---|---|
| δελφίς | dolphin |
| ἔγχελυς (f.) | eel |
| ἰχθύς | fish |
| κύων | shark (?) |
| κῆτος (n.) | sea–monster;[1] *Zauberross*;[2] whale? |
| πουλύπους | octopus |
| σπόγγος | sponge |
| τῆθος (n.) | sea–squirt |
| ὕδρος | water–snake (*Coluber natrix*) |
| φώκη | seal |

**Reptiles**

| | |
|---|---|
| δράκων | serpent |
| ὄφις | snake |

**Wild Beasts**

| | |
|---|---|
| αἴξ ἄγριος, or αἴξ ἀγρότερος | wild goat |
| ἄρκτος (f.) | bear |
| ἔλαφος (m., f.) | deer |
| νεβρός (m., f.) | fawn |
| ἐλλός (m.) | fawn |
| κέμας (f.) | young (2 yr.) deer[6] |
| πρόξ (f.) | roe[7] |
| θήρ | synonym for lion (usu. sing.); wild beasts (pl.) |
| θηρίον | wild beast |
| θώς (m., f.) | jackal |
| ἰκτίς | martin |

**Insects, etc.[3]**

| | |
|---|---|
| ἀκρίς | locust |

---

[1] D' A. W. Thompson, *A Glossary of Greek Fishes* (1947) 114.

[2] F. Schachermeyr, *Poseidon und die Entstehung des griechischen Götterglaubens* (1950) 200.

[3] See M. Davies and J. Kathirithamby, *Greek Insects* (1986), cf. G. Fernandez, *Nombres de insectos en griego antiguo* (1959).

---

[4] *Homerische Tierwelt*, 88.

[5] "μέλισσα" in G. Fernandez, *op. cit.* 62.

[6] *Homerische Tierwelt*, 50.

[7] LSJ.

| | | | |
|---|---|---|---|
| ἰκτίς | martin | σκύμνος | whelp |
| κνώδαλον | any wild or dangerous animal | λύκος | wolf |
| | | νυκτερίς (f.) | bat |
| | | πά(ο)ρδαλις (f.) | leopard |
| λαγωός[8] or πτώξ | hare | σῦς ἄγριος, or | wild boar |
| λέων | lion (=? *Leo, leo persicus*) | σῦς ἀγρότερος | wild boar |
| | | σῦς κάπριος | wild boar |
| θήρ | lion | κάπρος | wild boar |
| λίς | lion | | |

---

[8] subst. and adj.

# Appendix B
# Emotive States Attributed to Animals


There are approximately 70 words or phrases referring to seats of emotion or to psychological or mental states in animal similes. These are distributed among 82 of the animal similes, 60 of them developed. Therefore approximately 65% of the animal similes contain such vocabulary.[1] In 60% of the total 82 an analogous mental state is mentioned in the surrounding narrative. In non–animal similes, the concentration of emotive or ethical terms is less dense than in animal similes. Of the 43 non–animal similes containing emotive vocabulary, 35 different emotive terms may be discerned. Approximately 30% of the 125 non–animal similes, therefore, contain such vocabulary.

## Expressions and Words for Emotive States and Seats of Emotion Applied to Animals

### I. Courageous or Daring Spirit

φρῆν occurs in a variety of expressions and compounds, such as μέγα φρονέων, or ὀλοόφρων (7x), always with the boar or lion. Lion, 10.184, 486, 15.630, 17.111, 16.758; lion and boar together, 16.824; boar, 11.325.

σθένος occurs in a variety of expressions, such as σθένεϊ βλεμεαίνων with boar or lion. Lion, 17.135; boar or lion, 12.42, 5.783=7.257.

ἀλκή (cf. ἄλκιμος) occurs in a positive sense ten times altogether, of lion, boars, wolves, and wasps. Lions, 5.299, 17.61, 20.169, *Od.* 6.130, cf. 21.578 (leopard); boars, 4.253, 13.471, 17.281, 728; wolves, 16.157; wasps, 16.264. In a pejorative sense (i.e. preceded by οὐ- or ἀν-) with the following: fawn, 4.245, 13.104; lambs, 16.355.

---

[1] U. Dierauer, *Tier und Mensch im Denken der Antike, Studien zur antiken Philosophie* 6 (1977) 8, remarks on the frequency of similes in which the concord of human and animal drives and emotions is explicitly emphasized.

χαρμή occurs three times, once each in a positive sense with lion, 16.823, and eagle, 12.203 (omen); once negatively with fawn, 13.104.

ἦτορ: fawn, 11.115 (here as the physical organ)[2]; wasp, 16.264 (cf. ἄλκιμος above).

κῆρ: boar, 12.45.

θάρσος: flies, 17.570.

χόλος: lion 18.322, snake, 22.94.

ἀγήνωρ: lion, 12.300; cf. ἀγήνορι θυμῷ, 24.42.

## II. Expressions for Fear

τρόμος: deer, 11.117 (cf. ἄτρομος: wolf, 16.163).

φοβέομαι: cattle, sheep, 5.140, 11.172; smaller birds, 22.141.

δειδιότες: fish, 21.24.

κήδει: sheep, 17.550.

χλωρόν δέος αἱρεῖ: lion, 17.67 (cf. δείσας of the lion, *Od.* 4.792)

οὐ (δέ) ... τάρβει ... οὐδὲ φοβεῖται: boar or lion, 12.45f.; leopard, 21.574f.

## III. θυμός

The use of θυμός twice suggests a cognitive state: the θυμός bids the animal to attack: lion, 12.300; hawk, 22.142.

Four times θυμός refers to the exhaustion of the life–spirit of the animal: boar, 12.150; horse 16.468 (narr.), hare 17.678; fish, *Od.* 22.388.

Twice θυμός refers to the brawn of a domesticated animal: cows, 13.704; mules, 17.744.

θυμός refers to the courageous attributes of the wolf, 16.162 and leopard, 21.574.

Twice θυμός in the expression τετιηότι θυμῷ refers to disappointment in a lion, 11.555=17.664.

## IV. Volition

ἐθέλουσι, 16.825, and ἀέκων, 17.112, of the lion and boar, imply volition.

---

[2]But cf. K. von Fritz, "Νόος and νοεῖν in the Homeric Poems," *CPh* 38 (1943) 81, who argues that the emotive function and the organ are not distinguished.

## V.  Yearning

ποθέων/-οντες of the horse, 11.161, 24.6

## VI.  Joy

ἀγαλλόμενα of birds in flight, 2.462.

ἔχαρη of the lion, 3.23.

ἀγλάιηφι πεποιθώς of the horse, 6.510=15.267.

# Appendix C
# Formulae in Lion Similes

If we examine the language of the lion similes from a formular point of view, initially applying Parry's rigid concept of the formula,[1] and require a minimum of three occurrences for a phrase to achieve formulaic status, there is only one strictly formulaic lion simile: λείουσιν ἐοικότες ὠμοφάγοισι (ν) 5.782, 7.256, 15.592. It falls at end–line position. The introductory phrase ὡς δὲ λέων is found at initial position in the developed similes at 5.161, 10.485, 11.113. Four short formulaic phrases occur within lion similes:

a) ὥς τίς τε λέων occurs three times in Book 17 (133, 542, 657, cf. 61 ὡς δ' ὅτε τίς τε λέων). The repeated phrase coming after the first colon in the verse exhibits metrical regularity.

b) ἀπὸ μεσσαύλοιο at end–line position: 11.548, 17.112, 657.

c) λίς ἠυγένειος, containing the alternate term for lion, falls between the bucolic diaeresis and the end of the verse: 15.275, 17.109, 18.318.

d) λαβὼν κρατεροῖσιν ὀδοῦσιν at end–line position: 11.114, 11.175=17.63.

There are three instances of repetitions of entire hexameter lines in lion similes.

a) 11.550–55=17.659–64. The initial two lines of each simile differ, as do the resumptive clauses.

b) 11.175f.=17.63f. Both similes describe lions attacking and devouring a cow and share in common those lines conveying the essential actions: the method of attack and the lion's eating habits.

c)       δηρὸν ἔῃ κρειῶν, κέλεται δέ ἑ θυμὸς ἀγήνωρ       12.300f.
      μήλων πειρήσοντα καὶ ἐς πυκινὸν δόμον ἐλθεῖν·

---

[1] "Studies in the Epic Technique of Oral Verse–Making, I: Homer and Homeric Style," *HSCP* 41 (1930) 80 (=*MHV* 272).

cf.

> ἠὲ μετ᾽ ἀγροτέρας ἐλάφους· κέλεται δέ ἑ γαστὴρ *Od*. 6.133f
> μήλων πειρήσοντα καὶ ἐς πυκινὸν δόμον ἐλθεῖν·

In summary, there is one formulaic lion simile, a repeated phrase introducing three developed similes, four formulaic phrases recurring three times each within developed similes, and finally three blocks of repeated lines. All of the vocabulary is restricted.

Given the relatively small quantity of hexameter lines in lion similes, as well as the topical subject matter, one may be justified in reorienting the criteria for judging their formulaic character. If we apply to lion similes Hainsworth's concept of the formular modification as a repeated word–group (exhibiting flexibility in respect to word–order, separation of constituent words, insertion or omission of particles or prepositions, and inflexion)[2] we may gain a fairer impression of the language not already brought to light through an application of Parry's criteria. Except for Table 3 (phrases shared among lion similes and other animal similes), we shall concern ourselves only with vocabulary repeated a minimum or two times in lion similes. We may then observe that of those phrases restricted to lion similes, a total of seven recur at least once. Of those phrases and significant words occurring in lion similes as well as narrative, a total of ten recur within the body of lion similes. Finally, there are eleven phrases or words occurring in lion similes that are shared with other kinds of animal similes, notably boar similes. From the results of a detailed study of phraseology in lion similes we may conclude that the density of formulae in lion similes is higher than for any other type of animal simile, and that formulaic phrases occur in lion similes and the narrative of the *Iliad* with a comparable rate of frequency.[3]

One may note an increase in the amount of repeated vocabulary in lion similes when Hainsworth's modifications are employed. Given the similarity of setting from simile to simile and the special subject matter, instances of formulae restricted to lion similes are not surprising. In as large a body of similes as the lion similes one would expect to encounter repeated specialized phrases for the physical setting, the act of plundering the steading, the method of attack, the efforts to repulse the invader and the

---

[2] Hainsworth, *Flexibility* 36.

[3] Cf. W. B. Ingalls, "Formular Density in the Similes of the Iliad," *TAPA* 109 (1979) 87–109, who uses 2.87–94 to show that similes were composed orally.

lion's meal. (See Table 1.) The recurring phraseology assists the poet in staging a simile and also strengthens the cumulative visual impact of his simile type. In a type as well established as lion similes unexpected variations will be noticeable and can alert the audience to significant developments in the narrative.

Certain adjectives and phrases occurring in lion similes recur in other types of animal similes. The majority refer to heroic virtues or internal states (Table 2, a, b, e, h; Table 3, h, i, j); the correspondence between human and bestial appetites (Table 2, f; Table 3, a, b); and to flight and its opposite—qualities which the similes in general are wont to emphasize (Table 3, k). We may note a tendency towards vocabulary shared among lion and boar similes, where the beasts' opponents (men and dogs) are essentially the same: sc. κύνες θαλεροί τ' αἰζηοί (3.26, 11.414).

In the non–restricted vocabulary, i.e., that which is shared with narrative, one may observe once again the overlapping with the narrative of vocabulary relating to motion on the one hand and emotion on the other (Table 2, a–e; Table 3, e). The tactical maneuver of spear–throwing is common to warriors and farmers (Table 1, c).

**Tables Showing Repeated Phraseology from Lion Similes[4]**

**Table 1. Corresponding Phrases Restricted to Lion Similes**

a) πρῶτον, ἔπειτα δέ θ' αἷμα καὶ ἔγκατα πάντα
  λαφύσσει                                                    (11.176=17.64)
  cf. ἔγκατα καὶ μέλαν αἷμα λαφύσσετον                         (18.583)
  cf. ἔγκατά τε σάρκας τε καὶ ὀστέα μυελόεντα              (*Od.* 9.293)

b) τῆς δ' ἐξ αὐχέν' ἔαξε λαβὼν κρατεροῖσιν ὀδοῦσι(ν)
                                                             (11.175=17.63)
  cf. ῥηϊδίως συνέαξε λαβὼν κρατεροῖσιν ὀδοῦσιν              (11.114)
  cf. θορὼν ἐξ αὐχένα ἄξῃ                                      (5.161)

c) ... θαμέες γὰρ ἄκοντες                               (11.552f.=17.661f.)

---

[4] Except for Table 3 (phrases shared between lion similes and other animals similes), the Tables take into account phrases repeated a minimum of two times in lion similes. Words and phrases appearing in the resumptive clauses of similes are considered part of the simile.

cf. ἀντίον ἀίσσουσι θρασειάων ἀπὸ χειρῶν    (11.552=17.662)
cf. ἀντίον ἵστανται καὶ ἀκοντίζουσι θαμειὰς
   αἰχμὰς ἐκ χειρῶν...    (12.44f.)

d) ὅς τε σταθμοὺς κεραΐζων    (16.752)
cf. σταθμοὺς ἀνθρώπων κεραΐζετον    (5.557)

e) πρώτῃσι καὶ ὑστατίῃσι βόεσσιν    (15.634)
cf. πρώτῃσι βόεσσι    (18.579)

f) σταθμοῖο δίεσθαι    (12.304)
cf. σταθμοῖο δίωνται    (17.110)

g) λέων ὀρεσίτροφος    (12.299, 17.61, *Od.* 6.130, *Od.* 9.292)

## Table 2: Non–Restricted Repeated Lion Vocabulary

| Phrase in lion simile | Occurs elsewhere |
|---|---|
| a) θυμός ἐνὶ στήθεσσι(ν): 17.22, 68 | narrative:16x *Il.*, 16x *Od.* |
| b) ἄλκιμον ἦτορ: 17.111, 20.169 | narrative: 5.529, 16.209, 264 (wasp simile) |
| c) ἐντροπαλιζόμενος: 11.547, 17.109 | narrative: 6.496, 21.492 |
| d) ἄλλομαι and compounds: 5.142, 12.305, 16.754 | narrative: 12.466, 16.755, etc. |
| e) φοβέω: 5.140, 11.121, 172, 173, 178 | frequent in narrative |
| f) κέλεται δέ ἑ/με θυμὸς (ἀγήνωρ): 12.300 κέλεται δέ ἑ γαστήρ: *Od.* 6.133 | narrative: 10.534, 19.187 |
| g) πυκινὸν δόμον (ἐλθεῖν): 12.301, *Od.* 6.134 | narrative: 10.267 |
| h) θυμὸς ἀγήνωρ: 12.300 cf. ἀγήνορι θυμῷ: 24.42 ἀγηνορίη: 12.46 (lion–boar simile) | narrative: 8x *Il.* |
| i) ὅ(υ)ρεος/-ων (ἐν) κορυφῇσ(ι)/κορυφάς: 16.757, 824 | non–animal simile: 2.456, 3.10, 12.282 |

### Table 3: Lion Vocabulary Shared With Other Animal Similes

Eleven phrases or words occurring in lion similes are shared with other kinds of animal similes, notably boar similes. For purposes of clarification similes of an either–or (i.e., lion–boar [j, k]) nature are given at the end. An asterisk indicates that the word or phrase occurs in the narrative as well.

| Lion simile | Other simile |
|---|---|
| a) ὠμοφάγος (5.782, 7.256, 15.592) | jackal (11.479) |
| | wolf (16.157) |
| b) σίντης (20.165) | wolf (16.353) |
| c) *σμερδαλέος (18.579) | snake (22.95) |
| d) *αἴθων (10.178, 11.547, 18.161) | cattle, horses, eagle, and often in narrative |
| e) ἁρπάζω (5.556, 12.305, 13.199, 17.62) | eagle (22.310) |
| f) *ἀλκὶ πεποιθώς (5.299, 17.61) | boar (13.471, 17.728) |
| g) κύνες θαλεροί τ' αἰζηοί (3.26) | boar (11.414) |
| h) *ὀλοόφρων (15.630) | boar (17.21) |
| i) ταρβεῖ οὐδὲ φοβεῖται (12.46) | leopard (21.575) |
| j) *σθένεϊ βλεμεαίνει (17.22, 135) cf. σθένεϊ βλεμεαίνων | lion or boar (12.42) |
| k) *αἰὲν ἀποκτείνων τὸν ὀπίστατον· οἱ δὲ φέβοντο (11.178) | lion or boar (8.342) |

# Appendix D
## A List of Line ReferencesTo Lion Similes in Homer

[Including θήρ (marked by asterisk); marauding lion similes are given in
**bold-faced** type.]

### Developed Lion Similes

3.23–6 (4 lines)
**5.136–42** (7 lines)
**5.161–62** (2 lines)
**5.554–58** (5 lines)
**10.183–86*** (4 lines)
**10.485–86** (2 lines)
11.113–19 (7 lines)
**11.172–76** (5 lines)
**11.548–55** (8 lines)
12.41–49 (8 lines)
**12.299–306** (8 lines)
15.271–76 (6 lines)
**15.324–25*** (2 lines)
**15.586–88*** (3 lines)
15.630–36 (7 lines)
**16.487–89** (3 lines)
16.756–58 (3 lines)

16.823–26 (4 lines)
**17.61–67** (7 lines)
**17.109–12** (4 lines)
17.133–36 (4 lines)
**17.657–64** (8 lines)
18.317–22 (5 lines)
**18.573–86** (14 lines: Shield vignette)
**20.164–73** (10 lines)
22.262–64 (3 lines: oaths parable)
**24.41–43** (3 lines)
*Od*. 4.335–39 [=17.126–130] (4 lines each)
*Od*. 4.791–92 (2 lines)
***Od*. 6.130–34** (5 lines)
***Od*. 22.402–405** (4 lines)

### Short Lion Similes

3.449*
5.299
5.476
5.782 [=7.256=15.592]
110.297
11.129

11.239
11.383
11.546*
2.293
15.592
17.20

**17.542**
24.572
*Od*. 9.292
*Od*. 10.218
*Od*. 23.48

# Bibliography

A. Abramowiczowna, "Encore une fois la chasse au sanglier τ 393–466," *Eos* 68 (1980) 37–40.

E. Alston, and C. Danford, "On the Mammals of Asia Minor, II," *Proc. Zoo. Soc. Lond.* (1880) 50–66.

J. K. Anderson, *Ancient Greek Horsemanship* (Berkeley and Los Angeles 1961).

—— *Hunting in the Ancient World* (Berkeley, Los Angeles, and London 1985).

M. Andronikos, *Totenkult, Archaeologia Homerica* W (Göttingen 1968).

H. W. Auden, "Natural History in Homer," *CR* 10 (1896) 107.

(1973–4) 219–74.

N. Austin, "The One and the Many in the Homeric Cosmos," *Arion* 1/2 (1973–4) 219–74.

M. Baltes, "Zur Eigenart und Funktion von Gleichnissen im 16. Buch der *Ilias*," *A&A* 29 (1983) 36–48.

H. Bannert, "Zur Vogelgestalt der Götter bei Homer," *WS* N.F. 12 (1978) 29–42.

S. E. Bassett, "The Function of the Homeric Simile," *TAPA* 52 (1921) 132–47.

W. Beck, "Choice and Context: Metrical Doublets for Hera," *AJP* 107 (1986) 480–88.

F. Bechtel, *Lexilogus zu Homer* (Halle 1914; reprint, Hildesheim 1964).

F. Bernheim, & A. A. Zener, "The Sminthian Apollo and the Epidemic Among the Achaeans at Troy," *TAPA* 108 (1978) 11–14.

C. R. Beye, "Repeated Similes in the Homeric Poems," in *Studies Presented to Sterling Dow on his Eightieth Birthday, GRB* Monograph 10 (Durham, North Carolina) (1984) 7–13.

C. W. Blegen et al., *Troy*, 4 vols. (Princeton 1958).

E. Block, "The Narrator Speaks: Apostrophe in Homer and Virgil," *TAPA* 112 (1982) 7–22.

R. and E. Blum, *The Dangerous Hour. The Lore of Crisis and Mystery in Rural Greece* (London 1970).

J. Boardman, and D. Kurtz, *Greek Burial Practices* (London 1974).

J. Boessneck, and A. von den Driesch, "Ein Löweknochenfund aus Tiryns," *AA* 94 (1979) 447–49.

—— "Ein Beleg für das Vorkommen des Löwen auf der Peloponnes in 'herakleischer Zeit'," *AA* 96 (1981) 257–258.

E. Boisacq, *Dictionnaire étymologique de la langue grecque* (Paris 1923).

H. Bonjour, "Beowulf and the Beasts of Battle," *PMLA* 72 (1957) 563–73.

A. Bonnafé, "Quelques remarques à propos des comparaisons homériques de l' *Iliade*: critères de classification et étude statistique," *RPh* 57 (1983) 77–79.

—— "Eumée divin porcher et meneur d' hommes," *BAGB* 2 (1984) 180–94.

J. Borasten, "The Birds of Homer," *JHS* 31 (1911) 216–50.

C. M. Bowra, *Heroic Poetry* (London 1950).

—— *Tradition and Design in the Iliad* (Oxford 1930; reprint, Westport, Conn., 1977).

J. M. Bremer et al., ed., *Homer: Beyond Oral Poetry. Recent Trends in Homeric Interpretation* (Amsterdam 1987).

M. K. Brown, *Symbolic Lions. A Study in Ancient Mesopotamian Art and Literature*. (diss. Harvard 1974) Summary in *HSCP* 78 (1974) 273–75.

H.–G. Buchholz et al., *Jagd und Fischfang, Archaeologia Homerica* J (Göttingen 1973).

W. Burkert, *Homo Necans* (Berlin 1972); tr. Peter Bing (Berkeley, Los Angeles, and London 1983).

—— *Greek Religion*, (Stuttgart 1977), tr. J. Raffan (Cambridge, Mass. 1985).

R. W. Bushnell, "Reading 'Winged Words'. Homeric Bird Signs, Similes, and Epiphanies," *Helios* 9 (1982) 1–13.

H. Cahn, "Die Löwen des Apollon," *MH* 7 (1950) 185–99.

J. K. Campbell, *Honour, Family and Patronage. A Study of the Institutions and Moral Values in a Greek Mountain Community* (Oxford 1964).

R. Carpenter, *Folk Tale, Fiction and Saga in the Homeric Epic* (Berkeley 1958).

J. Caskey, *Lerna:* Vol. I: *The Fauna,* by Nils–G. Gejvall (Princeton 1969).

J. Chadwick and M. Ventris, *Documents in Mycenaean Greek²* (Cambridge 1973).

P. Chantraine, "Études sur le vocabulaire grec; quelques termes du vocabulaire pastoral et du vocabulaire de la chasse," *Études et Commentaires* 24 (1956) 31–96.

——— *Dictionnaire étymologique de la langue grecque. Histoire des mots* I–V (Paris 1968–80).

——— *Grammaire homérique, I: Phonétique et Morphologie²* (Paris 1973); *II: Syntaxe* (reprint, Paris 1986).

R. Chaplain, *The Study of Animal Bones from Archaeological Sites* (London 1971).

A. Clausing, *Kritik und Exegese der homerischen Gleichnisse im Altertum* (diss. Freiburg 1913).

M. Coffey, "The Function of the Homeric Simile," *AJP* 78 (1957) 113–32.

R. Conley et al., *European Hog Research Project,* Tennessee State Game & Fish Commission (Nashville, Tenn. 1973).

J. C. Cressey, "The Dogs of War. *Iliad* 22," *LCM* 7 (1982) 22–24.

J. E. Croft, *Pastoral Elements in the Iliad and Odyssey* (diss. Princeton 1973) Summary in *DA* 34 (1974) 7208A.

C. Darwin, *The Expression of the Emotions in Man and Animal* (London 1872).

M. Davies, and J. Kathirithamby, *Greek Insects* (Oxford 1986).

E. Delebecque, *Le cheval dans l' Iliade; Lexique du cheval chez Homère; Essai sur le cheval pré–homérique. Études et Commentaires* 9 (Paris 1951).

N. De Wit, *Le rôle et le sens du lion dans l' Égypte ancienne* (Paris 1951).

U. Dierauer, *Tier und Mensch im Denken der Antike, Studien zur antiken Philosophie* 6 (Amsterdam 1977).

B. C. Dietrich, "Xanthus' Prediction. A Memory of Popular Cult in Homer," *AClass* 7 (1964) 9–24.

F. Dirlmeier, *Die Vogelgestalt homerischer Götter*, *SHAW* 2 (Heidelberg 1967).

J. M. Duban, "Distortion as a Poetic Device in the 'Pursuit of Hektor' and Related Events," *Aevum* 54 (1980) 3–22.

J. du Boulay, *Portrait of a Greek Mountain Village* (Oxford 1970).

J. Duchemin, "Aspects pastoraux de la poésie homérique. Les comparaisons dans *l' Iliade*," *REG* 73 (1960) 362–415.

H. Dunbar, *A Complete Concordance to the Odyssey of Homer*, [ed. B. Marzullo] (New York 1971).

P. E. Easterling, "Notes on Tragedy and Epic," in: *Papers Given at a Colloquium on Greek Drama in Honour of R. P. Winnington–Ingram*, ed. by L. Rodley, Society for the Promotion of Hellenic Studies, Supplementary Paper 15 (London 1987) 52–62.

M. W. Edwards, *Homer, Poet of the Iliad* (Baltimore 1987).

H. Erbse, (ed.), *Scholia Graeca in Homeri Iliadem (Scholia vetera)*, I–V (+ index vols.) (Berlin 1969–1988).

—— "Homerische Götter in Vogelgestalt," *Hermes* 108 (1980) 259–74.

—— *Untersuchungen zur Funktion der Götter im homerischen Epos, Untersuchungen zur antiken Literatur und Geschichte* 24 (1986).

M. Faust, "Die künstlerische Verwendung von κύων, 'Hund' in den homerischen Epen," *Glotta* 48 (1970) 8–31.

B. Fenik, *Typical Battle Scenes in the Iliad, Hermes* Einzelschrift 21 (Wiesbaden 1968).

G. Fernandez, *Nombres de insectos en griego antiguo* (Madrid 1959).

M. I. Finley, "Marriage, Sale and Gift in the Homeric World," *Rev. Internat. des Droits de l' Antiquité*, 3rd ser. 2 (1955) 167–94.

——*The World of Odysseus*² (Harmondsworth 1978).

H. P. Foley, " 'Reverse Similes' and Sex Roles in the Odyssey," *Arethusa* 11 (1978) 14–21.

S. Foltiny, "The Ivory Horse Bits of Homer and the Bone Horse Bits of Reality," *BJ* 167 (1967) 11–37.

H. Fränkel, *Die homerischen Gleichnisse.*² *Mit einem Nachwort und einem Literaturverzeichnis*, ed. E. Heitsch (Göttingen 1977).

H. Fränkel, *Dichtung und Philosophie des frühen Griechentums* ² (1962); tr. M. Hadas and J. Willis (1975).

E. Friedl, *Vasilika. A Village in Modern Greece* (New York 1962).

R. Friedrich, "On the Compositional Use of Similes in the Odyssey," *AJP* 102 (1981) 120–37.

W. H. Friedrich, "Von homerischen Gleichnissen und ihren Schicksalen," *A&A* 28 (1982) 103–30.

K. von Fritz, "Νόος and νοεῖν in the Homeric Poems," *CPh* 38 (1943) 79–93.

H. Gabelmann *Studien zum frühgriechischen Löwenbild* (Berlin 1965).

J. Gonda, *Remarks on Similes in Sanskrit Literature* (Leiden 1949).

J. Griffin, *Homer on Life and Death* (Oxford 1980).

—— "Homeric Pathos and Objectivity," *CQ* 26 (1976) 161–87.

—— "The Epic Cycle and the Uniqueness of Homer," *JHS* 97 (1977) 39–53.

G. Guggisberg, *Simba, the Life of the Lion* (London 1961).

J. B. Hainsworth, "Odysseus and the Dogs," *G&R* 8 (1961) 122–25.

—— *The Flexibility of the Homeric Formula* (Oxford 1968).

—— *Homer. G&R New Surveys in the Classics*, 3 (Oxford 1969).

—— "Criticism of an Oral Homer," *JHS* 90 (1970) 90–8.

—— "Good and Bad Formulae in *Homer*," in *Tradition and Invention* , ed. B. Fenik (Leiden 1978) 41–50.

R. Hampe, *Die Gleichnisse Homers und die Bildkunst seiner Zeit* (Tübingen 1952).

K. Harder, [review of Körner *Homerische Tierwelt*] *Wochenschrift für kl. Phil.* (1918) 27.

F. M. Heichelheim, "Das Tier in der Vorstellungswelt der Griechen," *Studium Generale* 20 (1967) 85–89.

C. G. Heyne, *Homeri Carmina* (Leipzig 1802–22) 9 vols.

J. C. Hogan, *The Oral Nature of the Homeric Simile* (diss. Cornell 1966). Summary in *DA* 27 (1966) 1352A.

J. P. Holoka, "Homer Studies 1971–1979," *CW* 73 (1979) 65–150.

—— "'Looking Darkly' (ΥΠΟΔΡΑ ΙΔΩΝ): Reflections on Status and Decorum in Homer," *TAPA* 113 (1983).

F. Hölscher, *Die Bedeutung archaischer Tierkampfbilder, Beiträge zur Archäologie* V (Würzburg 1972).

D. B. Hull, *Hounds and Hunting in Ancient Greece* (London, Chicago 1964).

W. B. Ingalls, "The Analogical Formula in Homer," *TAPA* 106 (1976) 211–26.

—— "Formular Density in the Similes of the Iliad," *TAPA* 109 (1979) 87–109.

G. Jachmann, *Der homerische Schiffskatalog und die Ilias, Arbeitsgem. für Forsch. des Landes Nordrhein–Westfalen,* Wiss. Abh.5 (Cologne 1958), esp. Exkurs III, "Die homerische Gleichnisbildung," 267–338.

T. Jahn, *Zum Wortfeld "Seele–Geist" in der Sprache Homers, Zetemata* 83 (Munich 1987).

R. Janko, *Homer, Hesiod and the Hymns* (Cambridge 1982).

—— "The Shield of Heracles and the Legend of Cycnus," *CQ* (1986) 38–59.

I. J. F. de Jong, "Fokalisation und die Homerischen Gleichnisse," *Mnemosyne* 38 (1985) 257–70.

—— *Narrators and Focalizers. The Presentation of the Story in the Iliad* (Amsterdam 1987).

J. Kakridis, and S. G. Kapsomenos, "Das Wespengleichnis im Π der *Ilias,*" *Hermes* 88 (1960) 250–53.

R. Kannicht, "Poetry and Art. Homer and the Monuments Afresh," *CA* 1 (1982) 70–86 and plates 1–6.

A. Keith, *Simile and Metaphor in Greek Poetry from Homer to Aeschylus* (Menasha, Wisc., 1914 ).

O. Keller, *Die antike Tierwelt* (Leipzig 1909–13; reprint, Hildesheim 1963).

J. Kindstrand, "Homer in den Tiergeschichten des Ailianos," *Hermes* 104 (1976) 35–53.

G. S. Kirk, "Objective Dating Criteria in Homer," *MH* 17 (1960) 189–205.

—— *The Songs of Homer* (Cambridge 1962).

—— (ed.) *The Language and Background of Homer* (Cambridge 1964).

—— *Homer and the Oral Tradition* (Cambridge 1976).

G. S. Kirk, *The Iliad: A Commentary I: Books 1–4* (Cambridge 1985).

O. Körner, *Die homerische Tierwelt*[2] (Munich 1930).

—— *Das homerische Tiersystem und seine Bedeutung für die zoologische Systematik des Aristoteles* (Wiesbaden 1917).

T. Krischer, *Formale Konventionen der homerischen Epik*, Zetemata 56 (Munich 1971).

W. K. Lacey, "Homeric ἔδνα and Penelope's κύριος," *JHS* 86 (1966) 55–68.

J. Latacz, *Kampfparänese, Kampfdarstellung und Kampfwirklichkeit der Ilias, bei Kallinos und Tyrtaios.* (Munich 1977).

—— ed., *Homer. Tradition und Neuerung* (Darmstadt 1979).

R. Lattimore, (tr.) *The Iliad of Homer* (London 1951; Chicago 1951 & 1962).

—— (tr.) *The Odyssey of Homer* (New York 1965).

E. R. Leach, "Anthropological Aspects of Language: Animal Categories and Verbal Abuse," in *New Directions in the Study of Language,* [ed., E. H. Lenneberg] (Cambridge, Mass.) 1964) 23–63.

W. Leaf, *The Iliad*, 2 vols. (London 1900–02).

D. Lee, *The Similes of the Iliad and Odyssey Compared* (Melbourne 1966).

M. Leumann, *Homerische Wörter* (Basel 1950).

H. Liddell, R. Scott *A Greek–English Lexicon*[9], revised by H. S. Jones and R. McKenzie (Oxford 1940) with a supplement (1968).

S. Lilja, "Theriophily in Homer," *Arctos* 8 (1974) 71–8.

—— *Dogs in Ancient Greek Poetry* (Helsinki 1976).

B. Lincoln, "Homeric λύσσα: 'Wolfish Rage'," *IF* 80 (1975) 98–105.

S. H. Lonsdale, "Attitudes Towards Animals in Ancient Greece," *G & R* 26 (1979) 146–59.

—— "If Looks Could Kill: παπταίνω and Related Imagery in Homer," *CJ* 84 (1989) 325–333.

—— "Protean Forms and Disguise in *Odyssey* 4," *Lexis* 2 (1988) 165–78.

—— "Hesiod's Hawk and Nightingale (*Op.* 202–213): Fable or Omen?," *Hermes* 117 (1989) 403–12.

A. B. Lord, *Singer of Tales* (Cambridge, Mass. 1960).

A. B. Lord, "Homer as Oral Poet," *HSCP* 72 (1968).

G. Lorenz, *Die Einstellung der Griechen zum Tiere* (diss. Innsbruck 1972).

C. W. Macleod, *Homer. Iliad XXIV* (Cambridge 1982).

F. P. Magoun, "The Theme of the Beasts of Battle in Anglo–Saxon Poetry," *Neuphilologische Mitteilungen* 56 (1955) 81–90.

W. T. Magrath, "Progression of the Lion Simile in the *Odyssey*," *CJ* 72 (1982) 205–12.

C. Mainoldi, *L' Image du loup et du chien dans la Grèce ancienne d' Homère à Platon* (1984).

G. E. Markoe, "The 'Lion Attack' in Archaic Greek Art: Heroic Triumph," *CA* 8 (1989) 86–115.

E. Masson, *Recherches sur les plus anciens emprunts sémitiques en grec* (Paris 1967).

F. Matz, ed. *Archaeologia Homerica* (Wiesbaden 1967–).

M. McCall, *Ancient Rhetorical Theories of Simile and Comparison* (Cambridge, Mass. 1969).

L. Meule, "La faune d' Homère," *Soc. de Zoo. de Paris* (1910).

A. B. Meyer, "The Antiquity of Lions in Ancient Greece," *Annual Report of the Board of Regents*, Smithsonian Institution (1903) 661–67.

R. M. Meyer, *Deutsche Stilistik²* (Munich 1913).

W. Minton, "Invocation and Catalogue in Hesiod and Homer," *TAPA* 91 (1961) 22–30.

W. Moog, "Naturgleichnisse und Naturschilderungen bei Homer," *Zeitschrift für angewandte Psychologie* 6 (1912) 123–73.

I. Morris, "The Use and Abuse of Homer," *CA* 5 (1986) 81–138.

C. Moulton, "Similes in the *Iliad*," *Hermes* 102 (1974) 381–97.

—— *Similes in the Homeric Poems, Hypomnemata* 49 (Göttingen 1977).

—— "Homeric Metaphor," *CPh* 74 (1979) 279–93.

M. Mueller, *The Iliad* (London, Boston, and Sydney 1984).

M. Mühl, "Relikte der Tier– und Sachstrafe bei Homer," *REG* (1971) 1–16.

G. Mylonas, "The Lion in Mycenaean Times," *AAA* 3 (1970) 421–25.

J. L. Myres, "Homeric Art," *ABSA* 45 (1950) 229–60.

M. N. Nagler, *Spontaneity and Tradition* (Berkeley, Los Angeles, and London 1974).

G. Nagy, *The Best of the Achaeans* (Baltimore 1979).

M. P. Nilsson, *Homer and Mycenae* (London 1933).

—— *Geschichte der griechischen Religion³, Hdbch. d. Altertumswiss.* (Munich 1967).

M. C. Nussbaum, *The Fragility of Goodness* (Cambridge 1986).

—— *Aristotle' s* De Motu Animalium (Princeton 1978).

D. Offermanns, *Der Physiologus nach den Handschriften G und M* (Meisenheim am Glan 1966).

D. Ohly, *Griechische Goldenbleche des 8. Jahrhunderts v. Chr.* (Berlin 1953).

R. B. Onians, *The Origins of European Thought* (Cambridge 1951).

J. C. Opstelten, "Καλλίτριχες ἵπποι," *Hermeneus* 32 (1960) 77–79.

D. W. Packard et al., *A Bibliography of Homeric Scholarship* (Malibu 1974).

D. L. Page, *History and the Homeric Iliad* (Berkeley 1959).

E. Papamichael, "'Αχιλλεὺς ... ὠμηστὴς ἀνήρ," *Dodone* 9 (1980) 131–35.

A. Parry, "The Language of Achilles," *TAPA* 87 (1956) 1–7.

—— "Language and Characterization in Homer," *HSCP* 76 (1972) 9–22.

M. Parry, "The Distinctive Character of Enjambement in Homeric Verse," *TAPA* 60 (1929) 200–20.

—— "Studies in the Epic Technique of Oral Verse–making, I: Homer and Homeric style," *HSCP* 41 (1930) 73–147.

—— *The Making of Homeric Verse. The Collected Papers of Milman Parry*, edited with an introduction by A. Parry (Oxford 1971).

V. Pedrick, "The Paradigmatic Nature of Nestor's Speech in *Iliad* 11," *TAPA* 113 (1983) 55–68.

A. Persson, "Legende und Mythos in ihrem Verhältnis zu Bild und Gleichnis im vorgeschichtlichen Griechenland," in *DRAGMA Martino Nilsson* (Lund 1939) 379–401.

W. Peterson, "Some Greek Examples of Word–contamination," *AJP* 56 (1935) 54–60.

A. Podlecki, "Bird Omens in the Odyssey," *G&R* 14 (1967) 12–23.

—— "Some Odyssey Similes," *G&R* 18 (1971) 81–90.

J. Pollard, *Birds in Greek Life and Myth* (London 1977).

M. W. M. Pope, "The Parry–Lord Theory of Homeric Composition," *AClass* 6 (1963) 1–21; reprinted in *Homer. Tradition und Neuerung*, ed. J. Latacz (Darmstadt 1979) 338–67.

S. Porzig, *Die Namen für Satzinhalte im Griechischen und im Indogermanischen, Untersuchungen z. indogerm. Sprach– u. Kulturwiss.*, 10 (Berlin 1942).

H. Potratz, "Reiten und Fahren nach den Ergebnissen der Bodenforschung," *Prähistorische Zeit.* 30–31 (1939–40) 385–92.

G. Prendergast, *A Complete Concordance to the Iliad of Homer* [ed. B. Marzullo] (New York 1971).

R. A. Prier, "That Gaze of the Hound; *Odyssey* 19.228–321," *RhM* 123 (1980) 178–80.

H. Rahn, "Tier und Mensch in der homerischen Auffassung der Wirklichkeit. Ein Beitrag zur geisteswissenschaftlichen Selbstkritik," *Paideuma* 5 (1953) 277–97, 431–80 (reprint, Darmstadt 1968).

—— "Das Tier in der homerischen Dichtung," *Studium Generale* 20 (1967) 90–105.

G. Rapp and S. Aschenbrenner, eds. *Excavations at Nichoria in Southwest Greece* (Minneapolis 1978) I.

J. M. Redfield, *Nature and Culture in the Iliad. The Tragedy of Hektor* (Chicago 1975).

N. J. Richardson, "Literary Criticism in the Exegetical Scholia (bT) to the *Iliad*," *CQ* 30 (1980) 265–87.

W. Richter, *Die Landwirtschaft im homerischen Zeitalter, Archaeologia Homerica* H (Göttingen 1968).

S. Rocca, *Etiologia virgiliana, Pubblicazioni dell' Istituto di filologia classica e medievale* 80 (Genoa 1983).

H. Rohdich, "Der Hund Argos und die Anfänge des bürgerlichen Selbstbewusstseins," *A&A* 26 (1980) 33–50.

G. P. Rose, "Odysseus' Barking Heart," *TAPA* 109 (1979) 215–30.

C. J. Rowe, "Conceptions of Colour and Colour Symbolism in the Ancient World," *Eranos Jahrbuch* 41 (1972) 327–64.

J. Sachs, "Notes on Homeric Zoology," *TAPA* 17 (1886) 17–23.

I. Saunders, *The Rainbow in the Rock* (Cambridge, Mass., 1962).

F. Schachermeyr, *Poseidon und die Entstehung des griechischen Götterglaubens* (Munich 1950).

W. Schadewaldt, "Die homerische Gleichniswelt und die kretisch-mykenische Kunst," in Έρμηνεία, *Festschrift zum 60. Geburtstag O. Regenbogen dargebracht von Schülern und Freunden* (Heidelberg 1952) 9–27 (=*Von Homers Welt und Werk*[3] [Stuttgart 1965] 130–154).

—— "Die homerische Gleichniswelt und die kretisch–mykenische Kunst. Zur homerischen Naturanschauung," *Gymnasium* 60 (1953) 193–209.

S. Schein, *The Mortal Hero* (Berkeley, Los Angeles, and London 1984).

S. Shelmerdine, "Odyssean Allusions in the Fourth Homeric Hymn," *TAPA* 116 (1986) 49–64.

M. Schmidt, *Die Erklärungen zum Weltbild Homers und zur Kultur der Heroenzeit in den bT–Scholien zur Ilias, Zetemata* 62 (Munich 1976).

A. Schnapp–Gourbeillon, *L' animal dans Homère* (diss. Caen 1976).

—— *Lions, héros, masques. Les représentations de l' animal chez Homère* (Paris 1981).

—— "Le lion et le loup. Diomédie et Dolonie dans l' *Iliade*," *QS* 8 (1982) 45–77.

M. Schuster, "Bemerkungen zur homerischen Tierwelt," *MVPhW* (1928) 28–46.

W. C. Scott, *The Oral Nature of the Homeric Simile, Mnemosyne* Supp. 28 (Leiden 1974).

S. Scully, "The Language of Achilles: the ὀχθήσας Formulas," *TAPA* 114 (1984) 11–27.

C. Segal, *The Theme of the Mutilation of the Corpse, Mnemosyne* Supp. 27 (Leiden 1973).

—— "Le structuralisme et Homère: sauvageries, bestialité, et le problème d' Achille dans les derniers livres de l' *Iliade*," *Didactica Classica Gandensia* 17–18 (1977–78).

A. Severyns, "Simples remarques sur les comparaisons homériques," *BCH* (1946) 540–47.

H. Seyffert, *Die Gleichnisse der Odyssee* (diss. Kiel 1949).

G. P. Shipp, *Studies in the Language of Homer*[2] (Cambridge 1972).

M. S. Silk, *Interaction in Poetic Imagery* (Cambridge 1974).

B. Simon, *Mind and Madness in Ancient Greece* (Ithaca and London 1978).

B. Snell, *The Discovery of the Mind* [tr. T. G. Rosenmeyer] (Oxford 1953).

K. Snipes, "Literary Interpretation in the Homeric Scholia: The Similes of the *Iliad*," *AJP* 109 (1988) 196-222.

A. M. Snodgrass, "An Historical Homeric Society?" *JHS* 94 (1974) 114–125.

C. Spurgeon, *Shakespeare's Imagery* (Cambridge 1935).

W. B. Stanford, *Greek Metaphor* (Oxford 1936).

—— *The Odyssey of Homer, I: Books I–XII.* (London 1947); *II: Books XIII–XXIV* (London 1948).

G. Strasburger, *Die kleinen Kämpfer der Ilias* (diss. Frankfurt 1954).

W. Thalmann, *Conventions of Form and Thought in Early Greek Epic Poetry* (Baltimore 1984).

H. Thiry, "Homero y el perro de Hades. *Iliada* VIII y *Odisea* XI.623," *Emerita* 52 (1974) 103–108.

D' A. W. Thompson, *A Glossary of Greek Birds*[2] (Oxford 1936).

—— *A Glossary of Greek Fishes* (Oxford 1947).

A. Thornton, "The Aristeia of Diomedes as the Main Theme in the Composition of Books 5–8 of the *Iliad*," AULLA 18, Congress Proceedings (1977).

—— *Homer's Iliad : Its Composition and the Motif of Supplication, Hypomnemata* 81 (Göttingen 1984).

M. van der Valk, "Homer's Nationalistic Attitude," *AC* 22 (1953) 5–26.

A. Venkantasubbiah, "On Indra's Winning of Cows and Water," *Zeit. d. deutschen morgendländischen Gesellschaft* 115 (1965) 120–33.

N. Vereskshagin, *The Mammals of the Caucasus* (Jerusalem 1967).

E. T. Vermeule, *Götterkult, Archaeologia Homerica* V (Göttingen 1974).

—— *The Art of the Shaft Graves* (Cincinnati 1975).

—— *Aspects of Death in Early Greek Art and Poetry* (Berkeley, Los Angeles, and London 1979).

K. Vickery, *Food in Ancient Greece. Illinois Studies in the Social Sciences*, 20 (Urbana 1936).

P.Vidal–Naquet, "Valeurs religieuses et mythiques de la terre et du sacrifice dans l' Odyssée," *Annales* (ECS) 25 (1970) 1278–97; reprinted in *Le chasseur noir. Formes de pensées et formes de société dans le monde grec* (Paris 1983) 39–68.

E. Visser, *Homerische Versifikationstechnik, Europäische Hochschulschriften*, Reihe 15, *Kl. Sprachen und Literaturen*, Bd. 34 (Frankfurt 1987).

P. Vivante, *The Homeric Imagination* (Bloomington 1975).

—— *The Epithets in Homer* (New Haven 1982).

A. J. B. Wace, and F. H. Stubbings, *A Companion to Homer* (London 1962).

H. Wace, "Lions from Mycenae," in *Studi in onore L. Banti* (Rome 1965), 341–44.

P. Walcot, "Cattle Raiding, Heroic Tradition, and Ritual: The Greek Evidence," *History of Religions* 18 (1979) 326–51.

T. B. L. Webster, *From Mycenae to Homer* (London 1958).

J. Wiesner, "Fahren und Reiten im Alteuropa und im alten Orient," in *Der alte Orient* (Leipzig 1939; reprint Hildesheim 1971) Bd. 38, 2–4.

J. Weisweiler, "Vorindogermanische Schichten der irischen Heldensage," *Zeit. f. Celtische Philologie* 23 (1954) 27–28.

M. L. West, *Hesiod. Works and Days* (Oxford 1978).

J. Whaler, "Animal Simile in *Paradise Lost*," *PMLA* 47 (1932) 534–53.

C. H. Whitman, *Homer and the Heroic Tradition* (Cambridge 1958 & New York 1965).

M. M. Willcock, "Mythological Paradeigma in the Iliad," *CQ* 14 (1964) 141–54.

M. Woronoff, "Les chevaux de Troie," in *Mélanges E. Delebecque* (Paris 1983) 487–96.

N. Yalouris, "Athena als Herrin der Pferde," *MHDV* 7 (1950) 19–101.

A. Yoder, *Animal Analogy in Shakespeare's Character Portrayal* (New York 1947).

G. Zeuner, *A History of the Domestication of Animals* (London 1963).

# Index Locorum

2.308ff.: 113
2.455ff.: 108f.
3.16ff.: 50
4.243ff.: 12
5.133ff.: 51f.
5.160ff.: 51f.
5.554ff.: 54f.
6.506ff.: 34
8.247ff.: 113
9.524ff.: 123–125, 127
11.101ff.: 58f.
11.291ff.: 77
11.165ff.: 56f.
11.414ff.: 78f.
11.472ff.: 72-74
12.146ff.: 76
12.221f..: 113
12.299ff.: 35, 63–65
12.40ff.: 61f.
15.312ff.: 65–67
15.579ff.: 118
15.605ff.: 68
16.428ff: 2f.
16.752f.: 114f.
16.754: 12
16.756ff.: 114f.
16.823ff.: 114f.
17.19ff.: 36
17.60ff.: 79f.
17.66f.: 43f.
17.110: 79f.
17.282ff.: 81

17.725ff.: 76, 82f.
17.755ff.: 115
18.161ff.: 86f.
18.318ff.: 11, 89f., 93f.
18.520ff.: 121f.
18.573ff.: 121f.
20.164ff.: 2, 40–42, 88f., 122
21.12ff.: 111
21.22ff.: 111
21.573ff.: 37f.
22.29: 95f.
22.59ff.: 95–98
22.188ff.: 93f.
22.261ff.: 100f.
22.317ff.: 99f.
22.508ff.: 98
*Od.* 1.319ff.: 116
*Od.* 3.371f.: 116
*Od.* 4.791f.: 106
*Od.* 6.130ff.: 17, 35
*Od.* 9.292f.: 49
*Od.* 10.240: 127
*Od.* 10.408ff.: 126
*Od.* 16.215ff.: 126
Athen. 1.9: 28
Arist.: *HA* 629 27ᵇff.: 105f.
Hes., *Erga* 202ff.: 100
Hes., *Op.* 161ff.: 120
*Hymn to Aphrodite* 159: 44f.
Ps.–Hes., *Sc.* 367ff.: 106-108
Ver., *Aen.* VII.481f: 127
Xen., *Cyn.* 1.1ff.: 20